CONCRETE SURFACE ENGINEERING

Modern Concrete Technology Series

A series of books presenting the state-of-the-art in concrete technology

Series Editors
Arnon Bentur
Faculty of Civil and Environmental Engineering Technion-Israel Institute of Technology

Sidney Mindess
Department of Civil Engineering University of British Columbia

Steel Corrosion in Concrete
A. Bentur, S. Diamond & N. Berke   Hb: 978-0-419-22530-0

Durability of Concrete in Cold Climates
M. Pigeon & R. Pleau   Hb: 978-0-419-19260-2

Concrete in Hot Environments
I. Soroka   Hb: 978-0-419-15970-4

Concrete in the Maritime Environment
P. K. Mehta   Hb: 978-1-85166-622-5

CONCRETE SURFACE ENGINEERING

Benoit Bissonnette
Laval University, Canada

Luc Courard
University of Liège, Belgium

Andrzej Garbacz
Warsaw University of Technology, Poland

CRC Press
Taylor & Francis Group
Boca Raton London New York

CRC Press is an imprint of the
Taylor & Francis Group, an **informa** business

A SPON PRESS BOOK

CRC Press
Taylor & Francis Group
6000 Broken Sound Parkway NW, Suite 300
Boca Raton, FL 33487-2742

First issued in paperback 2017

© 2016 by Taylor & Francis Group, LLC
CRC Press is an imprint of Taylor & Francis Group, an Informa business

No claim to original U.S. Government works

ISBN-13: 978-1-4987-0488-5 (hbk)
ISBN-13: 978-1-138-74854-5 (pbk)

Visit the Taylor & Francis Web site at
http://www.taylorandfrancis.com

and the CRC Press Web site at
http://www.crcpress.com

Contents

Acknowledgments

The authors are grateful to professors Michel Pigeon (Laval University), Robert Degeimbre (University of Liège), and Lech Czarnecki (Warsaw University of Technology), who initiated them in construction materials science and taught them the fundamentals that are at the basis of concrete surface engineering.

The authors also acknowledge the contribution of the governments of Québec, Wallonia-Brussels, and Poland, which have supported for more than 15 years their collaborative research work through the Wallonia-Brussels-Québec and Wallonia-Brussels-Poland (MNiSW) Scientific Cooperation Programs.

Authors

Benoît Bissonnette, FACI, is professor of civil engineering at Laval University in Quebec City, Quebec, Canada. He has comanaged the NSERC Industrial Research Chair on Durable Repair and Optimized Maintenance of Concrete Infrastructure in the recent years and has been involved in a wide variety of research projects devoted to concrete repair. Dr. Bissonnette is a member of several technical committees of the American Concrete Institute (ACI C223 Shrinkage-compensating concrete; ACI C364 Rehabilitation), the International Union of Laboratories and Experts in Construction Materials, Systems and Structures (RILEM TC 181-EAS, TC 193-RLS), and the International Concrete Repair Institute (ICRI Repair Materials and Methods Committee). He authored or coauthored more than 125 peer-reviewed papers.

Luc Courard is professor of building materials at the University of Liège in Belgium. After completing his PhD work on concrete surface characterization in the late 1990s, he went to Laval University for a postdoctoral fellowship devoted to surface preparation of concrete prior to repair. His research activities are still dedicated to concrete surface characterization, as well as to innovative repair materials and supplementary cementitious materials. Dr. Courard is a member of the American Concrete Institute, the International Union of Laboratories and Experts in Construction Materials, Systems and Structures, and the Belgian Group of Concrete. He authored or coauthored more than 140 peer-reviewed papers.

Andrzej Garbacz is professor in the Department of Building Materials Engineering at the Warsaw University of Technology, Warsaw, Poland. Dr. Garbacz's research interests include concrete-polymer composites, repair materials, microstructure characterization, nondestructive testing (NDT) techniques for concrete characterization, and evaluation of repair materials. In 2005, he became a senior member of the International

Union of Laboratories and Experts in Construction Materials, Systems and Structures. He has been participating actively in the International Congress on Polymers in Concrete (ICPIC) since 1992, being appointed as a member of the board of directors in 2007 and as a secretary in 2013. He authored or coauthored more than 120 peer-reviewed papers.

Chapter 1

Introduction

1.1 FUNDAMENTALS OF SURFACE ENGINEERING

"Surface analysis, in conjunction with surface science and applied surface science, is a major activity contributing to our daily well-being, through improvements in energy, the environment, health, transport, consumer products and defense" (Briggs and Seah, 1990). The surface of a solid material interacts with the surrounding environment. This interaction can degrade the surface phase over time due to various phenomena, including wear, corrosion, fatigue, chemical reactions, and creep.

Surface engineering intends to characterize and improve the properties of the surface phase in order to prevent or slow down the degradation over time. This is accomplished by making the surface more resistant and robust to the environment in which it will be used.

Surface engineering spans a wide range of processes including plating technologies, nano and emerging technologies, characterization, and testing. At one end of the spectrum, ion implantation, nitriding, and aluminizing affect the chemistry and properties of only a thin surface layer of the substrate, by modifying the existing surface to a depth of 0.001–1.0 mm. At the other end of the spectrum are weld hard facings and other cladding processes: typically 1–20 mm thick, these are used for wear or corrosion resistance, as well as for repairing damaged elements.

Between these two extremes, a variety of deposition processes has been developed, such as physical vapor deposition (PVD), chemical vapor deposition (CVD), anodizing, laser processing, thermal spraying, cold spraying, and liquid deposition methods.

In the construction industry, anti-corrosion coatings for steel, coating protection of steel against corrosion or hydrophobic treatment for concrete are examples of surface treatments that contribute to preserving and even extending the service life of buildings and overall infrastructure by adapting the surface of materials to their environment. Recent research efforts in the field of surface engineering led to the development of breakthrough technologies such as self-cleaning glass, phase-changing materials for improved energetic performance, low-roughness pavement for noise reduction, and so on.

Surface engineering first developed into a legitimate discipline in the metallurgical field. According to Burakowski and Wierzchoń (1999), surface engineering of metals can be defined as a scientific and technological approach related to the design, the production, and the application of surface layers to improve some properties of the substrate, notably the resistance to corrosion and abrasion, as well as aesthetic properties. Many concepts of surface engineering, which developed into a discipline in the metallurgical field, can be applied to the surface treatment and repair of concrete structures. In the case of concrete surface engineering, it covers all relevant technical considerations from the surface characterization and its conditioning to the application of a new material that must adhere to it. Irrespective of the nature of the surface treatment or repair considered, suitable scientific tools are necessary to characterize the material properties, the characteristics of the substrate, and the adhesion developing between them. In this case, the implementation of surface engineering principles and tools is intended to produce a surface with enhanced characteristics to promote the development of both chemical and physical bonds, as the interfacial properties and stability are deeply influenced by surface preparation and conditioning. The primary purpose of this book is to sum up the body of knowledge in concrete surface engineering, based upon which sound practical approaches can be established for optimizing adhesion of repair and surface treatments for concrete.

1.2 CONCEPT AND SIGNIFICANCE OF SURFACE ENGINEERING APPLIED TO CONCRETE

When Long et al. (2001) asked, "Why assess the properties of near-to-surface concrete?," they were in touch with one of the most studied topics in the past decade. Repairing, coating or treating an existing concrete surface intrinsically involves potential interfacial problems between new and old materials. Compatibility assessment begins with the evaluation of the characteristics of the concrete substrate to be treated.

Concrete surface engineering is an area of knowledge addressing all surface-related considerations and their influence upon adhesion of repairs and other surface treatments (Courard and Garbacz, 2010). Many of the topics being addressed, for instance adhesion of polymer–concrete composites, are on the border—interface—between materials science and engineering. In any such composite system, the interface characteristics are necessarily influenced by the surface properties of the materials, such as surface roughness, wettability, viscosity, etc. Concrete surface engineering provides fundamental understanding of what will make the contact effective or not, allowing for interactions of variable intensity between the materials (Courard et al., 2009).

Implementation of a concrete surface engineering approach will contribute to improve the protection, maintenance and repair of concrete infrastructure, and, as a result, preserve or extend the expected service life. An approach of that type is formalized in the European Standard EN 1504 for the repair and protection of concrete structures affected or threatened by steel corrosion. Three main types of interventions towards concrete surface quality improvement are considered:

- Improvement of near-to-surface layer quality by hydrophobic treatment or impregnation.
- Removal of deteriorated concrete and repair with mortar.
- Application of adhesive coating to improve barrier properties.

Actually, the term concrete surface engineering is a neologism created to emphasize the need for a more fundamental approach of concrete repair engineering (Courard et al., 2009). Durability of repair and protection of concrete structures primarily depend on adhesion: favorable conditions during the development of bond between the substrate and the new material will guarantee its longevity and, consequently, of the repair. This is particularly important in the case of concrete–polymer composites: their naturally high adhesion can compensate for some characteristics less compatible with the existing concrete substrate (Czarnecki, 2008). Table 1.1 presents a list of parameters which may influence the creation and stability of the interface between a polymer composite and an existing concrete substrate.

The ability of two materials to bond and form a monolithic assembly or composite system is due to the creation of a continuous solid interface between these two materials (Kinloch, 1987): from a thermodynamic point of view, this means that the work of adhesion is greater than the work of cohesion. In an attempt to explain the fundamental causes of the phenomenon, one has to define and measure the electrical, molecular, and atomic forces existing between the two bonded materials and characterize the topography of the interface (Chapter 4). These phenomena define adhesion at the microscopic level, whereas adhesion measured by pull-off test is a gross macroscopic evaluation of the force (or energy) necessary to separate the bodies (Derjagin et al., 1978). In order to quantify these forces, different properties and characteristics of the two adhering meterials need to be evaluated. As an example, Table 1.1 presents a list of parameters which influence the creation and stability of interface between a new material and concrete substrate; specifically, parameters relating to the substrate are of particular interest in *concrete surface engineering*.

Obviously, the scope of interest of concrete surface engineering is not limited to the concrete surface itself: the approach is rather focused on generating sufficient adhesion between the two materials brought into intimate contact. Yet, the concept of adhesion has to be clearly defined because of

Table 1.1 Parameters and properties influencing the creation and stability of the interface

	Property	Symbol
Substrate		
Surface energy	Superficial tension	γ_S
Roughness	Effective surface waviness	r_f
Porosity	Water absorption	A
Capillary suction	Capillary absorption coefficient	A_{CA}
Saturation level	Water content	$[e]$
Mechanical characteristics	Superficial cohesion, E modulus	σ, E_S
Interstitial water	Chemical analysis	I
New layer		
Surface energy	Superficial tension	γ_L
	Interfacial tension	γ_{SL}
	Contact angle	θ
	Work of adhesion	W_A
	Effective bonding factor	ϕ_0
Binder setting	Setting	t_p
Kinetics of contact	Viscosity	η
	Binder content	E/C
	Time	t
	Temperature	T
Thermal dilatations	Thermal dilatation coefficient	α
Shrinkage	Water content	E/C
	Moisture	T, R.H.
	Geometrical characteristics	x
Porosity, capillarity	Water absorption	A
	Capillary absorption coefficient	A_{CA}
	Water vapor diffusion coefficient	μ
Mechanical characteristics	Cohesion	σ_A
	E modulus	E_A
	Stress release	R_A
Environment		
Hydrothermal conditions of substrate and air	Temperature, relative moisture	T, R.H.
Curing conditions	Temperature, moisture, protection	M
Human factors	—	F_H

Source: Courard, L., *Mater. Struct.*, 33, 65, 2000.

the "duality" of the term (Derjagin et al., 1978): "on one hand, adhesion is understood as a process through which two bodies are brought together and attached—bonded—to each other, in such a way that external force or thermal motion is required to break the bond. On the other hand, we can examine the process of breaking a bond between bodies that are already in contact. In this case, as a quantitative measure of the intensity of adhesion, we can take the force or the energy necessary to separate the two bodies."

Adhesion can thus be considered from different angles, depending on the phenomena being addressed: (1) the conditions and kinetics of contact or (2) the separation process. The intensity of adhesion will depend not only on the energy that is consumed to create the contact, but also on the interaction prevailing in the interfacial area (Courard, 2000). Generally speaking, two types of adhesion have to be considered: specific adhesion and mechanical interlocking (Figure 1.1).

Bond mechanisms depend on the true surface area, as opposed to the geometric surface area, and the contact surface area, also termed effective surface area (Figure 1.2). The contribution of mechanical adhesion (Derjagin et al., 1978) is explained by the fact that the liquid material penetrates through the roughness of the substrate and, after hardening,

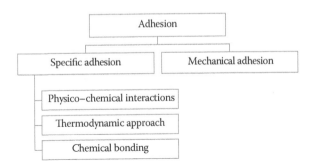

Figure 1.1 Principles of the theory of adhesion. (From Derjagin, B.V. et al., *Adhesion of Solids. Studies in Soviet Science: Physical Sciences*, Plenum Publishing Corporation, New York, 455pp., 1978.)

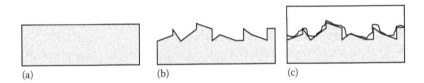

Figure 1.2 (a) Geometrical, (b) true, and (c) effective surface areas between substrate and overlay. (From Courard, L., *Mater. Struct.*, 33, 65, 2000.)

will produce adhesion by an interlocking effect. Special preparation technique (Chapter 6) can actually increase the true surface area (Courard and Darimont, 1998) and contribute to maximize the amount of potential interaction sites. It ought to be mentioned that the true surface area is difficult to measure, the evaluation being very sensitive to the measuring scale.

Ultimately, at the macroscopic scale, the level of adhesion generated between two materials depends on the actual surface where contact really exists between them, which is referred to as the *effective area* (Chapter 3).

Interfacial bond quality is the most important parameter in the success of a repair or surface treatment (Czarnecki, 2008). Quality is uneasy to define but, at the level of interface, it relies on all parameters and phenomena that may influence the nature and integrity of the bond between the concrete substrate and a repair material or surface treatment. In this respect, the overall compatibility between the two adhered materials (Chapter 5) appears to be of prime importance (Modjabi-Sangnier and Bissonnette, 2006). Over time, the various compatibility considerations (dimensional, permeability, chemical and electrochemical) govern the interactions between the two materials and, as a result, the evolution of interfacial bond integrity.

Bond strength is the macroscopic and measurable effect of these interactions. Quantification is usually made from pull-off test, shear test, or direct tensile test (Austin et al., 1995; Cleland and Long, 1997).

In different studies devoted to the problem of inadequate bond of repair materials and adhesives on existing concrete (Bijen and Salet, 1994; Fiebrich, 1994), the conclusions pointed at products and substances hindering adhesion of the repair or surface treatment material.

Pareek (1993) summarized the causes and cures for inadequate bond of repair systems (Table 1.2). At first, the classification of factors appears to be straighforward: on the one hand, there are factors related to the existing concrete substrate and, on the other hand, those related to the new material. Nevertheless, such a distinction is not possible for factors like thermodynamic properties or kinetics of contact, where interactions between the substrate and the new layer are aging (Courard and Darimont, 1998). It is important to keep in mind that the system is composed—from a thermodynamic point of view—of two bodies with their specific natures, but continuously affecting each other (Long et al., 2001). The interaction is quite complex as a result of the following phenomena:

- Aging character of the materials
- Porous structure of the materials
- Time-dependent deformations of the materials (shrinkage, creep)
- Variable temperature in time and space
- Variable moisture content in time and space

Table 1.2 Factors affecting adhesion of repair and finishing materials, and mitigation measures

Repair or finishing material	Substrate	Causes of poor adhesion to substrate	Damages	Improvement techniques
Ordinary cement mortar	• Cement concrete or mortar • Reinforcing bar	• Dry-out • Formation of voids at the interface • Drying shrinkage stresses at interface	• Loss of adhesion	Application of polymer dispersion, polymer-modified paste as bonding agent to substrate
Polymer-modified mortar	• Cement concrete or mortar • Reinforcing bar	• Improper type of polymer dispersion and polymer–cement ratio	• Cracking • Delamination	Selection of suitable surface treatment for substrate, and of appropriate type of polymer dispersion and polymer–cement ratio
Polymer mortar	• Cement concrete or mortar	• Wetness of substrate • Setting shrinkage of polymer mortar		Application of coupling agent as primer to substrate, and use of shrinkage-reduced polymer mortar

Source: Pareek, S.N., Improvement in adhesion of polymeric repair and finish materials for reinforced concrete structures, PhD thesis, Nihon University, Tokyo, Japan, 1993.

Moreover, one must never forget another category of factors, less tangible, but still fundamental for the quality of the work: human factors. It is not really possible to quantify them precisely, because they are often random and unpredictable, but it is important to take them into account.

There is evidence from these considerations that concrete surface engineering is essential for implementing successful concrete repair and surface treatment on a systematic basis. Adequate surface preparation (Chapter 6) and surface treatment (Chapter 7) are necessary for developing lasting bond. Another obvious finding at this point is that scientists and technicians involved in the evaluation, design, construction, maintenance and/or management of concrete infrastructure need to gain a better knowledge of the concrete surface make up, its properties and characterization, and understand the fundamentals and technical requirements of bond development.

REFERENCES

Austin, S., Robins, P., and Pan, Y. (1995) Tensile bond testing of concrete repairs. *Materials and Structures*, **28**(179), 249–259.

Bijen, J. and Salet, T. (1994) Adherence of young concrete to old concrete development of tools in civil engineering. In: *Proceedings of the Second Bolomey Workshop on Adherence of Young and Old Concrete*, Unterengstringen, Suisse, Switzerland (Ed. F.H. Wittman), Aedificatio Verlag, Freiburg, Germany, pp. 1–24.

Briggs, D. and Seah, M.P. (1990) *Practical Surface Analysis*. Vol. 1: *Auger and X-Ray Photoelectron Spectroscopy*, 2nd edn. Chichester, England: Wiley.

Burakowski, T. and Wierzchoń, T. (1999) *Surface Engineering of Metals: Principles, Equipment, Technology*. Boca Raton, FL: CRC Press.

Cleland, D.J. and Long, A.E. (1997) The pull-off test for concrete path repairs. *Proceedings of the Institution of Civil Engineers, Structures and Buildings*, pp. 451–460.

Courard, L. (2000) Parametric study for the creation of the interface between concrete and repair products. *Materials and Structures*, **33**, 65–72.

Courard, L. and Darimont, A. (1998) Appetency and adhesion: Analysis of the kinetics of contact between concrete and repairing mortars. In: *Proceedings of the International RILEM Conference on The Interfacial Transition Zone in Cementitious Composites*, Vol. 35 (Ed. A. Katz et al.), E&S Spon, London, U.K., pp. 207–215.

Courard, L. and Garbacz, A. (2010) Surfology: What does it mean for polymer concrete composites? *Restoration of Buildings and Monuments*, **16**(4/5), 291–302.

Courard, L., Michel, F., Schwall, D., Van der Wielen, A., Garbacz, A., Piotrowski, T., Perez, F., and Bissonnette, B. (2009) Surfology: Concrete substrate evaluation prior to repair. In: *Materials Characterization: Computational Methods and Experiments IV*, June 17–19 (Eds. A. Mammoli and C.A. Brebbia). The New Forest, U.K.: Wessex Institute of Technology Press, pp. 407–416.

Czarnecki, L. (2008) Adhesion—A challenge for concrete repair. In: *Proceedings of the Second International Conference on Concrete Repair, Rehabilitation and Retrofitting (ICCRRR'2008)*, Cape Town, South Africa (Eds. M.G. Alexander et al.), Taylor & Francis Group, London, U.K., 2009, pp. 935–940.

Derjagin, B.V., Krotova, N.A., and Smilga, V.P. (1978) *Adhesion of Solids. Studies in Soviet Science: Physical Sciences*. New York: Plenum Publishing Corporation, 455p.

Fiebrich, M.H. (1994) Scientific aspects of adhesion phenomena in the interface mineral substrate-polymers. In: *Proceedings of the Second Bolomey Workshop on Adherence of Young and Old Concrete*, Unterengstringen, Switzerland (Ed. F.H. Wittman), Aedificatio Verlag, Freiburg, Germany, pp. 25–58.

Kinloch, A.J. (1987) *Adhesion and Adhesives: Science and Technology*. London, U.K.: Chapman & Hall, 441pp.

Long, A.E., Henderson, G.D., and Montgomery, F.R. (2001) Why assess the properties of near-to-surface concrete? *Construction and Building Materials*, **15**, 65–79.

Modjabi-Sangnier, F. and Bissonnette, B. (2006) Durabilité et compatibilité des reparations en béton auto-nivellant. *Journées Scientifiques du (RF)²B*, Toulouse, France, Juin 19–20, 2007, 10pp. (in French).

Pareek, S.N. (1993) Improvement in adhesion of polymeric repair and finish materials for reinforced concrete structures. PhD thesis, Nihon University, Tokyo, Japan.

Pareek, S.N., Ohama, Y., and Demura, K. (1990) Adhesion mechanism of ordinary cement mortar to mortar substrates by polymer dispersion coatings. *Proceedings of the Sixth International Congress on Polymers in Concrete (ICPIC'90)*, Shanghai, China, pp. 442–449.

Chapter 2

Surface of concrete

2.1 SURFACE OF CONCRETE: CONCEPTS AND DEFINITIONS

The concrete surface layer is often referred to as its "skin." According to Kreijger (1984), three types of concrete skin can be defined: the cement skin (about 0.1 mm thick), the mortar skin (about 5 mm thick), and the concrete skin (about 30 mm thick), as a result of the wall effect, sedimentation and segregation, compacting methods, and water bleeding. In terms of appearance and aesthetical appreciation, the "concrete skin" would be corresponding to a depth of one-half the maximum aggregate size (Dieryck et al., 2005), as shown in Figure 2.1.

The skin of any cementitious material can in fact be defined as the surface layer where the effective particle size distribution is affected by the wall effect (formed surface) or the finishing operations (finished surface). Away from the skin, bulk concrete becomes more regular and less porous: it corresponds to concrete volume where crystals develop with less restriction.

In a broader sense, in this book, the near-to-surface layer (NSL) corresponds to the surface layer, irrespective of its depth and nature, whose chemical and physical characteristics will dictate its appearance, its interaction with the environment (Courard, 1999), and the applicability of a surface treatment system (paint, penetrating sealer, coating, repair material, etc.).

In any concrete maintenance or repair process, the concrete surface will be modified by surface preparation (Cleland et al., 1992). Irrespective of the surface preparation, the surface properties and characteristics will affect bond development (Courard and Garbacz, 2010). This is the reason it is fundamental to assess the substrate that comes into contact with the new material, more specifically its NSL (Garbacz et al., 2013). This interface—but we should more definitely consider an *interphase*—is the border between two phases (Figure 2.2): the substrate and the new material. If a clear "gray line" may be considered before contact (Figure 2.2a), it becomes less evident when contact is created: the two phases A and B will be mixed and interconnected (Figure 2.2b).

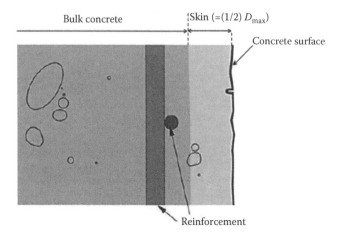

Bulk concrete

Skin (=(1/2) D_{max})

Concrete surface

Reinforcement

Figure 2.1 Concrete skin characteristics. (From Courard, L. et al., *Mater. Struct.*, 45(9), 1331, 2012.)

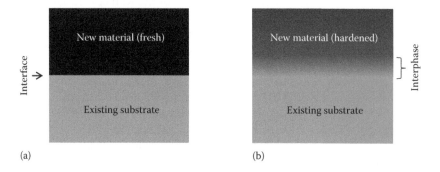

New material (fresh)

Interface →

Existing substrate

(a)

New material (hardened)

Interphase

Existing substrate

(b)

Figure 2.2 Interfacial zone in a treated concrete surface: (a) at the time of placement of the new material on the existing substrate and (b) after hardening of the new material. (From authors.)

The quality, specifically aesthetics, of concrete mainly depends on the chemical and physical properties of the concrete skin at a microscopic level (Lallemant et al., 2000). Parameters are numerous, including type and quality of materials, w/cm, environmental conditions of working, casting technology, curing, etc., and also formwork material (Courard et al., 2012).

2.2 CONCRETE SURFACE COMPOSITION

2.2.1 Formed concrete surface

The structure of NSL is considerably dependent on the way of designing and placing concrete. Conditions of curing will greatly influence the quality of concrete surface: bleeding or plastic shrinkage are phenomena appearing when fresh concrete is under unfavorable environmental conditions like heat, low humidity, or wind. The period of time (some hours to some days) when concrete is no more deformable but not enough resistant is particularly tremendous (Figure 2.3).

Demolding should consider one of the most important properties of the skin of the concrete—its color. This depends on many parameters (Table 2.1), proper action on which may guarantee a homogeneous aspect.

When concrete is cast into the formwork, phenomena similar to what happens in interfacial transition zone (ITZ) (see Section 2.2.3) may occur at the interface between concrete and form: the development of crystals—specifically tobermorite—is disturbed. Higher content of water, however,

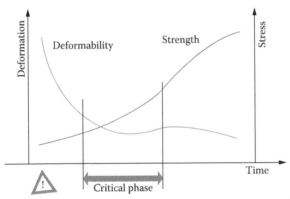

Decreasing strain limit and low early strength = increased risk of cracking

Figure 2.3 Critical period of time for early-age concrete. (From Groupement Belge du Béton, *Concrete Technology*, Belgian Federation of Cement, Brussels, Belgium, 582pp., 2010, in French.)

Table 2.1 Factors influencing concrete color

Parameter	More light	More dark
Portland cement	High ferrous oxide content	Low ferrous oxide content
GBBS cement	High GGBS content	Low GGBS content
W/C report	High	Low
Form material	Not absorbant	Absorbant
Form surface	Smooth	Rough
Time for demolding	Short	Long
Lime exudation	Large	Small
Evolution with time	Young	Old

Source: Groupement Belge du Béton, *Concrete Technology*, Belgian Federation of Cement, Brussels, Belgium, 582pp., 2010, in French.

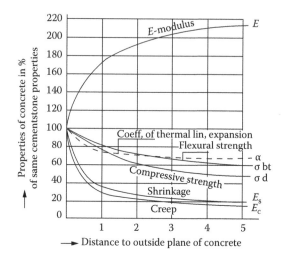

Figure 2.4 Estimation of concrete properties over the outside 5 mm of concrete surface. (From Kreijger, P.C., *Mater. Struct.*, 17(100), 275, 1984.)

allows the growth of large crystals of $Ca(OH)_2$, inducing higher porosity. This is also a reason for lighter surface vs. bulk concrete (Lemaire et al., 2005).

Mechanical performances are directly related to the microstructure of the NSL. Investigations (Kreijger, 1984) have revealed that E modulus is particularly low at the level of the NSL while compressive and bending strengths are higher than in bulk concrete (Figure 2.4).

Evaluation of physical properties also indicates that the concrete skin depth is 2 or 3 mm from the surface (Figure 2.5).

In some cases, it is possible to use textile specially designed for increasing the quality of the NSL (Figure 2.6): a decrease of the W/C ratio is generally observed. This effect is favorable not only for the quality of the skin—less porous, more resistant—but also for the durability of the structure itself by decreasing the rate of ingress of external aggressive products.

2.2.2 Original finished concrete surface

2.2.2.1 Usual finishing considerations

In the case of finished concrete surfaces, typically horizontal elements such as floors and pavements, the operations of floating/troweling and curing, in particular their respective timing, can play a crucial role in the formation and characteristics of the NSL.

The operation of concrete finishing in horizontal pours will typically involve floating and troweling. The purpose of floating (Figure 2.7) is

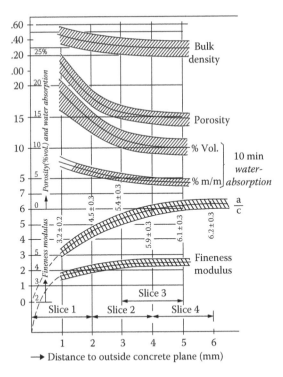

Figure 2.5 Composition and properties of concrete skin. (From Kreijger, P.C., *Mater. Struct.*, 17(100), 275, 1984.)

threefold: (1) to embed aggregate particles just beneath the surface; (2) to remove slight imperfections, humps, and voids; and (3) to compact the mortar at the surface in preparation for additional finishing operations. The concrete should not be overworked as this may bring an excess of water and fine material to the surface and result in subsequent surface defects. Where a smooth, hard, dense surface is desired, floating is normally followed by steel troweling. The floating/troweling operations need to be delayed until the concrete has hardened sufficiently so that water and fine material are not brought to the surface. Timing is very important: too long a delay will generally result in a surface that is too hard to float and trowel, while premature floating and troweling can cause scaling, crazing, or dusting and a surface with reduced wear resistance.

Exterior concrete should not be troweled, as it can lead to a loss of entrained air caused by overworking the surface. Floating and brooming can provide the nonslip surface texture needed for exterior concrete slabs.

The floating and troweling operations influence the thickness of the NSL, as it directly contributes in pushing the coarser particles below the surface. In addition, the timing of these operations with respect to the bleed water

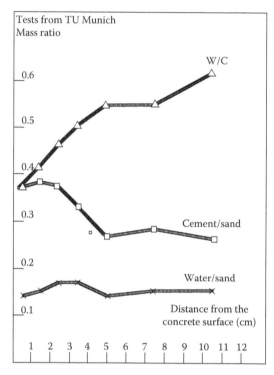

Tests from TU Munich
Mass ratio

Figure 2.6 Effect of form sheeting (Zemdrain™) on surface properties of concrete. (From DuPont de Nemours International S.A., Zemdrain controlled permeability formwork liner for longer life concrete, Technical Sheet, n.d., Le Grand-Saconnex, Switzerland.)

Figure 2.7 Power floating of a concrete ice rink floor with a ride-on equipment. (From www.essexyouthhockey.org.)

rise is extremely important, as it can affect considerably the effective water/cement ratio and thus the overall quality (porosity, strength, durability) of the paste in the NSL.

Immediately after finishing, all newly placed concrete slabs (and any other finished concrete) should be cured and protected from drying, from extreme changes in temperature, and from damage by subsequent construction and traffic. Curing is needed to ensure continued hydration of the cement, strength gain of the concrete, and a minimum of early drying shrinkage. Special precautions are necessary when concrete work continues during periods of adverse weather. In cold weather, arrangements should be made in advance for heating, covering, insulating, or enclosing the concrete. Hot-weather work may require special precautions against rapid evaporation and drying and high temperatures.

2.2.2.2 Architectural finishing considerations

Special procedures are implemented to obtain a variety of architectural concrete finish, such as

- Stamped concrete
- Exposed-aggregate surface
- Colored concrete

A variety of patterns and textures can be used to produce decorative finishes. Patterns can be formed with divider strips or by scoring or stamping the surface just before the concrete hardens. Textures can be produced with floats, trowels, and brooms, while more elaborate textures can be achieved with special techniques (CAC, 2011).

An exposed-aggregate finish provides a rugged, attractive surface in a wide range of textures and colors. In washing and brushing, the surface layer of mortar should be carefully washed away with a light spray of water and brushed until the desired exposure is achieved. The methods for exposing the aggregate typically include washing and brushing, using setting retarders, and scrubbing. When the concrete has hardened sufficiently, simultaneously brushing and flushing with water expose removes more or less the concrete skin, leaving the aggregate exposed (Figure 2.8).

The aggregates in exposed-aggregate concrete can also be exposed by mechanical means, such as waterblasting, bushhammering, grinding, and polishing.

Colored concrete finishes for decorative effects (Miller, 2013) can be achieved in a variety of applications by introducing pigments in the mixture (at the time of mixing) or applying a *dry shake* at the surface upon finishing.

In all earlier mentioned cases, the surface concrete is necessarily modified and generally different in composition and microstructure from the bulk concrete.

Figure 2.8 Exposed-aggregate concrete. (From Anonymous, www.pettitconcretefloor-ing.co.nz, accessed on March 31, 2015.)

2.2.3 Altered (treated) concrete surface

Surface treatment will generally remove the NSL and offer a new "skin" to the concrete.

Techniques induce low mechanical treatment of the surface by means of sand flow or mechanical shocks. They are applied when concrete is hardened, mainly for aesthetic purposes (Figure 2.9):

- Sandblasting
- Hammering
- Chiseling
- Polishing
- Diamond grinding

Depending on the type of equipment, techniques are used for effectively removing deteriorated or contaminated concrete. Thickness of removed concrete depends on the technique itself (Table 2.2).

2.2.4 Analogy with interfacial transition zone

The NSL due to wall effect offers analogies with the zone around aggregates in bulk concrete (namely, *Interfacial Transition Zone*) but also with the situation created at the interface between old concrete and new (repair) layer: in each of these situations, the "interphase" that has been created is different from the surrounding mortar or concrete.

(a)

(b)

(c)

(d)

(e)

(f)

Figure 2.9 Effect of different surface treatments on hardened concrete surface: (a) Sandblasting (From Anonymous, http://www.uniqrete.ie, accessed on March 31, 2015), (b) extended sandblasting (From Anonymous, http://www. atlanta-structural.net, accessed on March 31, 2015), (c) bush hammering (From Anonymous, http://www.hardscape.co.uk/products-colour.php?colour_id=2, accessed on March 31, 2015), (d) chiseling (From Anonymous, http://mger-wingarch.com/2010/08/29/architects-pet-peeve-no-14-concrete-vs-cement/, accessed on March 31, 2015), (e) grinding (From Anonymous, http://mpmex-tremefloors.com.au/extreme-floor-finishes-3/, accessed on March 31, 2015), and (f) grinding and polishing (From Anonymous, http://indecorativeconcrete. com/?page_id=855, accessed on March 31, 2015).

Table 2.2 Concrete removal tools and profile amplitude

Technique	Profile
Hand chipping	1/2 in. (13 mm)
Hydro milling	1/2 in. (13 mm)
Rotary milling	1/8–1/4 in. (3–6 mm)
Scabbling	1/8 in. (3 mm)
Shotblasting	1/8 in. (3 mm)
Abrasive blasting	1/8 in. (3 mm)

Source: Adapted from Bissonnette, B. et al., *Concrete Int.*, 28(12), 49, 2006.

Maso (1980) was the first to study the transition halo around the aggregates. His groundbreaking work evidenced that the transition zone is primarily due to a wall effect that induces gradients of W/C and to differences in mobility of the anhydrous cement particle ionic species. It is necessarily richer in hydrated constituents made of mobile ions in comparison with that of the bulk mass (Mehta and Monteiro, 1987). For Portland cements, ettringite and portlandite are the most common crystals present in this zone. The crystals are generally larger and better formed. Finally, at all ages, its porosity, ITZ's porosity (within a few μm from the surface of the aggregate) is always higher than in the bulk cement paste (Figure 2.10).

The thickness of the transition zone is typically of the order of 50 μm (Chatterji and Jensen, 1992). The greatest differences with regard to the properties are found in the first 20 μm. Moreover, the most sensitive area of the interface very often does not necessarily correspond to the "physical" interface (dimensionless) but is down 5–10 μm into the cement paste: this is due to preferential rupture planes in cleavage crystals of $Ca(OH)_2$

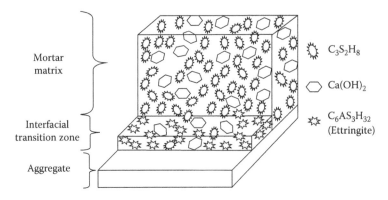

Figure 2.10 Representation of the interfacial transition zone (ITZ). (From Langton, C.A. and Roy, D.M., Morphology and microstructure of cement paste/rock interfacial regions. In: *Proceedings of the 7th International Congress on the Chemistry of Cement*, Vol. III, Septima edn., Paris, France, 1980, pp. VI-127–VII-132.)

(Mindess, 1987; Wang, 1987). The type of aggregate, aggregate roughness, and addition of silica fume are examples of parameters that appear to influence the orientation of the crystals of $Ca(OH)_2$ and, consequently, the intensity of the surface energies involved.

Van der Waal's forces play an important role in the interactions between materials. It is the density of the electron clouds of fluctuating atoms that induces dipoles enabling interaction between neighbour atoms and their polarization. The intensity of the interaction energy between two identical atoms or molecules is given by the following equation:

$$\varepsilon_d = \frac{3/4 h v_0 \alpha^2}{4 \pi \varepsilon x^6}$$

where

x is the distance between the atoms or molecules

$h v_0$ is the energy corresponding to the ionization potential of the atom or of the molecule

α is the potential of atoms or molecules to be polarized

The energy varies as a function of $1/x^6$, which means that when the distance between atoms is doubled, the intensity of the binding forces is 64 times less.

At the level of the hydrated cement phase, three types of bonds may occur: with tobermorite CSH or portlandite $Ca(OH)_2$ and inside $Ca(OH)_2$ and CSH. Silicate and sulfate anions are combined to lime by ionic bonding (Struble, 1987). Weaker bonding is generated between CSH and between CSH and other materials, the order of magnitude being in the range typical of hydrogen bridges and Van der Waal's forces (Tabor, 1981).

Observations made with a scanning electron microscope typically show the extensive presence of $Ca(OH)_2$ films, which are intrinsically characterized by a high porosity (Figure 2.11). These films are generally 30–50 μm

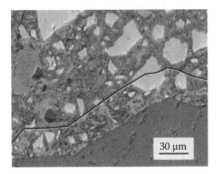

30 μm

Figure 2.11 View of the interfacial transition zone. (From Diamond, S. and Jingdong, H., *Cement Concr. Compos.*, 23(2–3), 179, 2001.)

thick after several days of hydration and may contain, in the case of contact with limestone aggregates, complexes of carboaluminates or calcium hydroxide–calcium carbonate (Odler and Zurz, 1987).

It is interesting, in the light of these considerations, to use the the paste-aggregate interface analogy in analyzing the application of a repair material or surface treatment on existing concrete: there are many similitudes, starting with the "wall effect". The analysis of fracture surfaces after tensile testing provides further information on the subject. Nonetheless, it must be stressed that the distances between the materials within a mixture or brought into contact with each other may be of different orders of magnitude.

When fresh cement paste comes into contact with a porous susbstrate like concrete, part of the liquid phase—water—is drawn into the pores of the substrate (Courard and Darimont, 1998). The consequence of this suction phenomenon is a decrease in the water content of the cement paste and, in case of excessive suction, the lack of water in the vicinity of the interface. A diffusion effect is also observed in addition to the phenomenon of capillary suction: as the most mobile ions—calcium, sodium, potassium, sulfate, and aluminum—are moving into the solution, the portlandite and ettringite concentrations increase at the interface (Figure 2.12). $Ca(OH)_2$ may also get carbonated into $CaCO_3$(Tam et al., 2005), although this happens to a much lesser extent than in the event of contact with a nonporous material.

When considering the contact between fresh cement paste and an existing concrete support, there is actually contact with cement paste (mortar) and aggregates respectively. This means substrate areas with very different ranges of porosity, usual concrete aggregates being much less porous than cement paste.

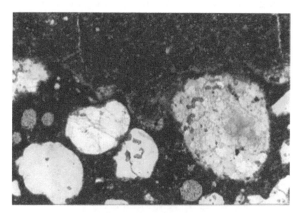

Figure 2.12 Interface between cement paste and concrete. (From Courard, L., Contribution à l'analyse des paramètres influençant la création de l'interface entre un béton et un système de reparation, Doctoral thesis, Faculty of Applied Sciences, University of Liege, Liege, Belgium, 1999, 213pp., in French.)

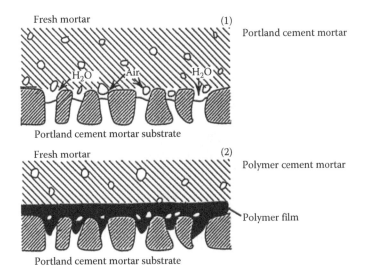

Figure 2.13 Interface between cement concrete (1) or polymer cement concrete (2) and concrete. (From Ohama, Y., *ACI Mater. J.*, 84(6), 511, 1987.)

Due to the microstructure of the existing concrete and the relative distance the forces acting at the interface between new and old materials are mainly of Van der Waal's type. There is in fact little chemical bond between concrete and new layer. Even when polymers are used (Figure 2.13), virtually no chemical bonds are created (Courard et al., 2014).

2.2.5 Surface concrete defects

Different types of degradation can be found on concrete structures such as leakage, settlements, deflection, wear, spalling, disintegration, scaling, delamination, etc. (Emmons and Vaysburd, 1994). The deterioration processes result from a variety of interacting physical, chemical and mechanical phenomena. Water is most generally involved in these processes, either directly (i.e., frost deterioration) or indirectly as the main vector for transport mechanisms (i.e., chloride penetration) inside the bulk concrete (Figure 2.14).

The causes of degradation may be classified into three main categories (Maage, 2004):

1. Causes of defects due to inadequate construction or materials
   - Inadequate structural design
   - Inadequate mix design, insufficient compaction, insufficient mixing
   - Insufficient cover
   - Insufficient or defective waterproofing
   - Contamination, poor or reactive aggregates
   - Inadequate curing

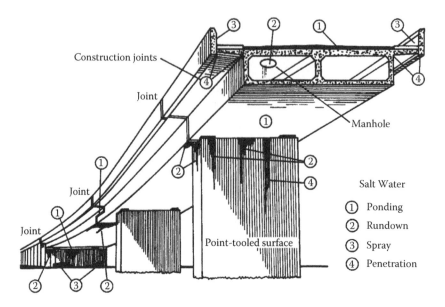

Construction joints

Joint

Manhole

Joint

Joint

Joint

Point-tooled surface

Salt Water

① Ponding
② Rundown
③ Spray
④ Penetration

Figure 2.14 Typical zones of chloride ions attack. (From Pritchard, B., *Bridge Design for Economy and Durability*, Thomas Telford Services, London, U.K., 1992, 172pp.)

2. Causes of defects revealed during service
   • Foundation movement, impacted movement joints, overloading
   • Impact damage, expansion forces from fires
3. External environment and agents (Figures 2.15 through 2.18)
   • Severe climate, atmospheric pollution, chloride, carbon dioxide, aggressive chemicals
   • Erosion, aggressive groundwater, seismic action
   • Stray electric currents

A general classification of the different causes of concrete degradation is presented in Figure 2.19. These degradations affect the quality of the NSL to a variable degree. The investigations necessary to qualify and quantify the defects and the causes of these defects must be organized in such a way that the origin of degradation can be identified and eventually acted upon (Schrader, 1992a,b,c).

Starting as early as the time of casting, different factors can affect the NSL. How concrete sets and hardens and the resulting surface characteristics depend on a number of parameters. In addition to the structure safety requirements, formwork, scaffolding, and other supporting structures should be such that the construction loads do not induce at any time excessive deformations that may result in concrete cracking.

Figure 2.15 Contamination of concrete bridge deck. (Photo courtesy of A. Darimont.)

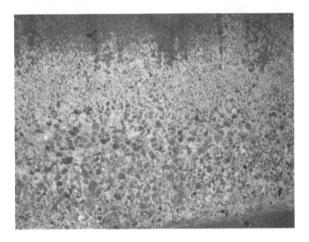

Figure 2.16 Effect on deicing salts—scaling—on bridge concrete protection structures. (From authors.)

It should also be provided sufficient counter deflection (taking into account the deformation of the formwork during casting, as well as the counter deflection required when the structure is to be put in service).

Provisions must be made for the formwork not to restrain the shrinkage of concrete, creating cracks before stripping. It is also important at the

Figure 2.17 Leakage of $Ca(OH)_2$ through cracked concrete bridge deck. (Photo courtesy of A. Darimont.)

Figure 2.18 Inefficient concrete beam repair operation. (From Emmons, P.H. and Vaysburd, A.M., *Constr. Build. Mater.*, 8(1), 5, 1994.)

design stage to favor geometrical characteristics less prone to shrinkage cracking.

The main defects and causes of disturbance in the NSL are the following (Figure 2.20):

- Difference of tint
- Honeycomb pattern
- Sandy zones
- Blowholes
- Impurities
- Cracks

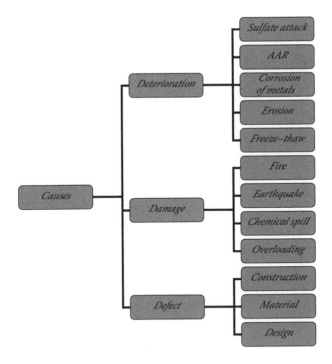

Figure 2.19 Causes of concrete degradation.

Many cracks, like those caused by restrained concrete shrinkage, are relatively shallow and are caused by forces that stabilize with time and do not lead in general to structural problems. Others, like those caused by foundation subsidence or changes in soil volume, are caused by driving forces that may remain active and result in serious structural problems over the long term.

The causes of cracking can be summarized as follows:

- Properties of the constituent materials
- Mixture design
- Surrounding environment to which concrete is exposed
- Mixing, placement, finishing, and curing practices
- Type of use
- Maintenance practices

Pattern cracking, also referred to as map cracking or crazing (Figure 2.21), appears as a network of random cracks at the surface of concrete. This type of cracking is usually shallow (less than 1/8 in. deep) and does not pose structural issue. Generally, it does not cause durability problems either, but it certainly may affect the aesthetics of the structure.

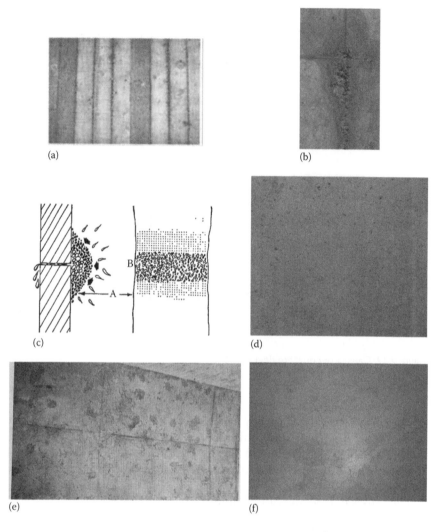

Figure 2.20 Different defects of concrete skin: (a) framework differential absorption, (b) honeycomb pattern, (c) sandy zone (B) due to cement migration (A), (d) blowholes, (e) impurities, and (f) cracks induced by plastic shrinkage. (From authors.)

Pattern cracking can be caused by the following:

• Curing practices inadequate for environmental conditions such as low humidity, high outside temperatures, direct sunlight, and wind, which can create high rates of evaporation from the surface layer of concrete; resistance to shrinkage from the underlying concrete causes stress that is relieved by surface crazing

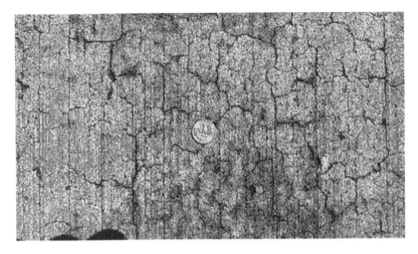

Figure 2.21 Pattern cracking. (From Anonymous, Visual Inspection of Concrete, InterNACHI, http://www.nachi.org/visual-inspection-concrete, accessed on March 31, 2015.)

- Excessive water in the mix
- Overvibration of the concrete, causing coarse aggregate to settle and cement paste to concentrate at the surface
- Overworking the surface with a steel trowel during finishing
- Performing finishing operations while bleed water is still on the surface
- Alkali aggregate reaction inducing expansion and gel exudation.
- Improper selection and implementation of surface preparation technique before repair (see Section 5.2)
- Sprinkling cement dust on the surface to soak up bleed water

2.3 EFFECTIVE COMPOSITION OF THE NEAR-TO-SURFACE LAYER

Aesthetic considerations as well as the need for achieving strong adhesion require comprehensive knowledge of the concrete surface composition. Fiebrich (1994) proposed an efficiency rating scale for concrete surface preparation based on the amount of exposed aggregates. The proposed scale was appraised through bond strength conducted on overlaid test specimens that were prepared with the use of high-pressure water jetting (900 bar). For grade A rated (less than 10 % of exposed aggregates) substrate, he recorded more than 2/3 of the failures at the interface with low bond strength values (1.2 MPa). Conversely, for concrete substrates rated with higher grades (from 30 to 80% of exposed aggregates), the number of failures recorded at the interface went down significantly (44%), while the average bond strength increased by 50% (1.8 MPa).

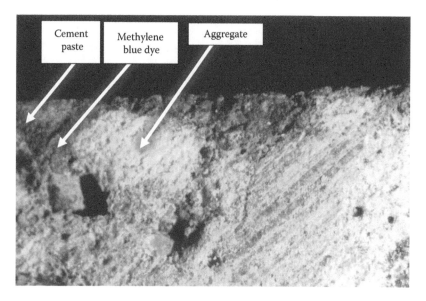

Figure 2.22 Penetration of methylene blue dye in a concrete substrate, between aggregates and cement paste. (From authors.)

Hence, when repairing or applying a treatment on concrete, the surface composition has clearly a major impact on the bond creation and development. For proper understanding and description of the mechanisms, it is necessary to go beyond the theoretical surface composition (Courard, 2005). As observed on Figure 2.22 (Courard, 1999), when applied on the concrete substrate, the new material penetrates through the porosity of the superficial layer, notably in the *Interfacial Transition Zone* around the exposed aggregates, more porous than the bulk paste (Tabor, 1981; Larbi and Bijen, 1991). Typically, the ITZ width in common grade Portland cement concrete (see Section 2.2.4) is of the order 50 µm (Maso, 1980) and, depending on the water-to-cementitious weight ratio, may reach up to 200 µm.

In studying the bond mechanisms between repair material or surface treatment and concrete, it is necessary to take into account the higher porosity around each aggregate particle, which actually enables deeper penetration of the new material. By extending the effective surface of contact between concrete substrate and repair material, this contributes to increase locally the anchorage of the new material, between bulk cement paste and aggregates (Courard and Degeimbre, 2003). Overall, given the relative importance of the cumulative ITZ's areas across the surface, it exerts a quite significant influence and the resulting bond strength is considerably higher than what could be expected based upon the theoretical composition of the concrete substrate surface (bulk paste and aggregates).

2.4 CONCLUSION

Development of adhesion in any concrete repair or surface treatment depends heavily on the characteristics of the near-to-surface layer of the existing concrete substrate. As in the *Interfacial Transition Zone* around the aggregates in concrete or in the superficial layer of a formed concrete surface, specific products will grow at the interface between the newly applied material and the substrate. The composition and topography of the NSL influence the characteristics of this interface and the bond being generated between the two adjoined materials. Hence, the mechanical, physical, and chemical properties of the receiving concrete surface need to be assessed in order to evaluate the ability of a new material to adhere satisfactorily. The following chapter is devoted to the techniques and test methods available for that matter.

REFERENCES

Bissonnette, B., Courard, L., Vaysburd, A., and Bélair, N. (2006) Concrete removal techniques: Influence on residual cracking and bond strength. *Concrete International*, 28(12), 49–55.

CAC. (2011) *Design and Control of Concrete Mixtures*, 8th Canadian edn., Canada's Cement Industry, Ottawa.

Cavins, J. (1999) Enhancements in surface quality of concrete through use of controlled permeability formwork liners. *Magazine of Concrete Research*, 51(2), 73–86.

Chatterji, S. and Jensen, A.D. (1992) Formation and development of interfacial zones between aggregates and Portland cement pastes in cement-based materials. In: *Proceedings of the RILEM International Conference on Interfaces in Cementitious Composites*, Toulouse, France (Ed. J.C. Maso), E&FN Spon, London, U.K., pp. 3–12.

Cleland, D.J., Yeoh, K.M., and Long, A.E. (1992) The influence of surface preparation methods on the adhesion strength of patch repairs for concrete. *Proceedings of the Third Colloquium on Materials Science and Restoration*, Esslingen, Germany, pp. 858–871.

Courard, L. (1999) Contribution à l'analyse des paramètres influençant la création de l'interface entre un béton et un système de réparation. Doctoral thesis, Faculty of Applied Sciences, University of Liege, Liege, Belgium, 213pp. (in French).

Courard, L. (2005) Adhesion of repair systems to concrete: Influence of interfacial topography and transport phenomena. *Magazine of Concrete Research*, 57(5), 273–282.

Courard, L. and Darimont, A. (1998) Appetency and adhesion: Analysis of the kinetics of contact between concrete and repairing mortars. In: *Proceedings of the RILEM International Conference on Interfacial Transition Zone in Cementitious Composites* (Eds. A. Katz, A. Bentur, M. Alexander, and G. Arliguie), London, U.K., E&FN Spon, Haïfa, Israel, pp. 185–194.

Courard, L. and Degeimbre, R. (2003) A capillary suction test for a better knowl-edge of adhesion process in repair technology. *Canadian Journal of Civil Engineering*, 30(6), 1101–1110.

Courard, L. and Garbacz, A. (2010) Surfology: What does it mean for polymer con-crete composites? *Restoration of Buildings and Monuments*, 16(4/5), 291–302.

Courard, L., Martin, M., Goffinet, C., Migeotte, N., Piérard, J., and Polet, V. (2012) Influence of the reuse of OSB and marine plywood formworks on con-crete facing color. *Materials and Structures*, 45(9), 1331–1343.

Courard, L., Piotrowski, T., and Garbacz, A. (2014) Near-to-surface properties affecting bond strength in concrete repair. *Cement and Concrete Composites*, 46, 73–80.

Diamond, S. and Jingdong, H. (2001) The ITZ in concrete—A different view based on image analysis and SEM observations. *Cement and Concrete Composites*, 23(2–3), 179–188.

Dieryck, V., Desmyter, J., Michel, F., and Courard, L. (2005) Surface quality of self-compacting concrete and raw materials properties. In: *Second North American Conference on the Design and Use of Self-Consolidating Concrete (SCC) and the Fourth International RILEM Symposium on Self-Compacting Concrete*, Chicago, IL (Ed. S.P. Shah), Northwestern University, Evanston, IL, pp. 287–295.

DuPont de Nemours International S.A. (n.d.) Zemdrain controlled permeability formwork liner for longer life concrete, Technical Sheet, Le Grand-Saconnex, Switzerland.

Emmons, P.H. and Vaysburd, A.M. (1994) Factors affecting the durability of concrete repair: The contractor's viewpoint. *Construction and Building Materials*, 8(1), 5–16.

Fiebrich, M.H. (1994) Scientific aspects of adhesion phenomena in the inter-face mineral substrate-polymers. In: *Proceedings of the Second Bolomey Workshop on Adherence of Young and Old Concrete* (Ed. F.H. Wittman), Aedificatio Verlag, Unterengstringen, Sion, Switzerland, pp. 25–58.

Garbacz, A., Courard, L., and Bissonnette, B. (2013) A surface engineering approach applicable to concrete repair engineering. *Bulletin of the Polish Academy of Sciences (Technical Sciences)*, 61(1), 73–84.

Groupement Belge du Béton. (2010) *Concrete Technology*. Belgian Federation of Cement, Brussels, Belgium, 582pp (in French).

Kreijger, P.C. (1984) The skin of concrete: Composition and properties. *Materials and Structures*, 17(100), 275–283.

Lallemant, I., Rougeau, P., Gallias, J.L., and Cabrillac, R. (2000) Contribution of microscopy to the characterization of concrete surfaces presenting local tint defects. *22nd International Conference on Cement Microscopy*, Montréal, Quebec, Canada, pp. 107–121.

Langton, C.A. and Roy, D.M. (1980) Morphology and microstructure of cement paste/rock interfacial regions. In: *Proceedings of the 7th International Congress on the Chemistry of Cement*, Vol. III, Septima edn., Paris, France, pp. VI-127–VII-132.

Larbi, J. and Bijen, J.M. (1991) The role of the cement paste-aggregate interfacial zone on water absorption and diffusion of ions and gases in concrete. In: *The Cement Paste Aggregate Interfacial Zone in Concrete* (Ed. J.M. Bijen), Technische Universiteit Delft, the Netherlands, pp. 76–93.

Lemaire, G., Escadeillas, G., and Ringot, E. (2005) Evaluating concrete surfaces using an image analysis process. *Construction and Building Materials*, **19**, 604–611.

Maage, M. (2004) The new European EN 1504 standard. Guidelines for consultant. *NORECON Seminar*, Kopenhagen, Denmark.

Maso, J.-C. (1980) Bonding between aggregates and hydrated cement paste. *Proceedings of the VIIth International Congress on the Chemistry of Cements*, Vol. 18, Septima edn., Paris, France, pp. 61–64 (in French).

Mehta, P.K. and Monteiro, J.M. (1987) Effect of aggregate, cement and mineral admixtures on the microstructure of the transition zone. In: *Bonding in Cementitious Composites* (Eds. S. Mindess and S. Shah), Materials Research Society, Pittsburgh, PA, pp. 65–75.

Miller, S.H. (2013) A design alternative for floor art. *Concrete International*, **35**(3), 60–62.

Mindess, S. (1987) Bonding in cementitious composites: How important is it? In: *Bonding in Cementitious Composites* (Eds. S. Mindess and S. Shah), Materials Research Society, Pittsburgh, PA, pp. 3–10.

Odler, I. and Zurz, A. (1987) Structure and bond strength of cement-aggregate interfaces. In: *Bonding in Cementitious Composites* (Eds. S. Mindess and S. Shah), Materials Research Society, Pittsburgh, PA, pp. 21–27.

Ohama, Y. (1987) Principle of latex modification and some typical properties of latex modified mortars and concretes. *ACI Materials Journal*, **84**(6), 511–518.

Pritchard, B. (1992) *Bridge Design for Economy and Durability*. London, U.K.: Thomas Telford Services, 172pp.

Schrader, E.K. (1992a) Mistakes, misconceptions, and controversial issues concerning concrete and concrete repairs (part I). *Concrete International*, **14**(9), 52–56.

Schrader, E.K. (1992b) Mistakes, misconceptions, and controversial issues concerning concrete and concrete repairs (parts II). *Concrete International*, **14**(10), 48–52.

Schrader, E.K. (1992c) Mistakes, misconceptions, and controversial issues concerning concrete and concrete repairs (parts III). *Concrete International*, **14**(11), 54–59.

Silfwerbrand, J. (1990) Improving concrete bond in repair bridge decks. *Concrete International*, **12**(9), 61–66.

Struble, L. (1987) Microstructure and fracture at the cement paste-aggregate interface. In: *Bonding in Cementitious Composites* (Eds. S. Mindess and S. Shah), Materials Research Society, Pittsburgh, PA, pp. 11–20.

Tabor, D. (1981) Principles of adhesion—Bonding in cement and concrete. In: *Adhesion Problems in the Recycling of Concrete* (Ed. P. Kreijger), Nato Scientific Affairs Division, New York, pp. 63–90.

Tam, V.W.Y., Gao, X.F., and Tam C.M. (2005) Carbonation around near aggregate regions of old hardened concrete cement paste. *Cement and Concrete Research*, **35**(6), 1180–1186.

Wang, J. (1987) Mechanism of orientation of $Ca(OH_2)$ crystals in interface layer between paste and aggregates in systems containing silica frame. In: *Bonding in Cementitious Composites* (Eds. S. Mindess and S. Shah), Materials Research Society, Pittsburgh, PA, pp. 127–132.

Chapter 3

Characterization of a concrete surface

3.1 INTRODUCTION

At the beginning of a project where repair, protection, or reinforcement works are planned on a concrete structure, a condition evaluation must be performed. The investigation must be thorough enough to get a reliable picture of the manifestation, extent, severity, causes, and effects of all repertoriated degradation mechanisms and to assess their progress and potential effects on the safety, serviceability, and service life of the structure.

Engineers need guidance on both the techniques available for the condition assessment of structures and the methods for data interpretation. Two investigation stages are usually involved in the process (Robery, 1995). The first stage consists of a rapid-scan visual assessment, often including limited sampling in areas obviously damaged, from which areas of interest can be selected for more detailed investigations. The second stage entails a detailed diagnostic survey that relies on selected destructive and/or nondestructive testing techniques.

The diagnosis of the concrete structure must be performed (Courard et al., 2006a,b) in view of the application of a repair product/system or an overlay for specific purposes or objectives such as aesthetics, resistance to abrasion, waterproofing performance, etc. Two main aspects have to be verified to assess correctly the substrate: the in-place mechanical properties of the concrete and its overall condition (NIT 216, 2000). A complete appraisal should be made of the defects in the concrete structure, their causes, and whether they affect the ability of the concrete structure to perform its function (Teodoru et al., 1991).

For repair methods involving the application of mortar and concrete, the European Standard EN 1504-10 recommends (Table 3.1) a series of investigations before and/or after preparation of the concrete substrate. Minimum requirements before protection and repair are given in the EN 1504 series; nonetheless, it is not intended to serve as an inspection guide for assessing the condition of the concrete structure before, during, and after repair (Maage, 2004).

Table 3.1 Summary of tests and observations for quality control in the case of patch repair

Test number	Characteristics	Test method or observation	Standards	Frequency	Status
1	Delamination	Hammer sounding (T)	EN 12504-2	Once before application	A
2	Cleanliness	Visual (O) Wipe test (T)		After preparation and immediately before application	A
3	Roughness	Visual sand test (O) or Profile meter (T)	ISO 3274		B
4	Surface tensile strength of substrate	Pull-off test (T)	EN 1542 BS 1881 BS 207		B
5	Crack movement	Mechanical or electrical gauges (O)	BS 1881 BS 206		C
6	Vibration	Accelerometer (O)			C
7	Temperature of the substrate	Thermometer (O)		Throughout application	A
8	Carbonation	Phenolphthalein test (T)	EN 104865		C
9	Chloride content	Site sampling and chemical analysis (T)	BS 1181-124 DB 5423.78 NTBUILD 208		C
10	Penetration of other contaminants	Site sampling and chemical analysis (T)			C
11	Electrical resistivity	Wenner test (T)			C
12	Compressive strength	Core and crushing test (T)—Rebound hammer test (T)	EN 12504-1 EN 12504-2		B

Source: EN 1504-10:2003, Products and systems for the protection and repair of concrete structures - definitions, requirements, quality control and evaluation of conformity - Part 10: Site application of products and systems and quality control of the works.

T, test; O, observation; A, for all intended uses; B, for certain intended uses where required by the specific or operating conditions; C, for special applications.

Three parameters have to be particularly well controlled before any repair work or coating application: roughness, cleanliness, and soundness of the substrate. Attention should notably be paid to curing products, delaminations, oil and grease, dust, and laitance. Moreover, some substrates can be deeply polluted by chemical products.

3.2 SURFACE PROFILE

3.2.1 General characteristics of the evaluation techniques applicable to concrete surfaces

Surface texture is the repetitive or random deviation from the nominal surface that forms the three-dimensional topography of the surface (Bhushan, 2001). A very general typology of a solid surface is presented in Figure 3.1. With respect to this classification, depending on the surface preparation technique used and the scale being considered, the concrete surface profile yielded can be either homogeneous or inhomogeneous. Therefore, considering the wide range of possibilities, surface texture characterization is somehow complex and the available methods are not applicable in all situations.

The roughness of the substrate is one of the parameters often considered to affect adhesion strength between repair material and existing concrete. Nevertheless, this has been controversial for a number of years; some reported bond test results have shown that surface roughness exerts only a minor influence on the tensile bond. For instance, in the tests performed by Silfwerbrand (1990), adhesion to rough, water-jetted surfaces was compared with bond to smooth, sandblasted surfaces. It was concluded that there could be a roughness "threshold value" beyond which further improvement of the substrate roughness would not enhance bond strength. According to these test results, the "threshold value" ought to be close to

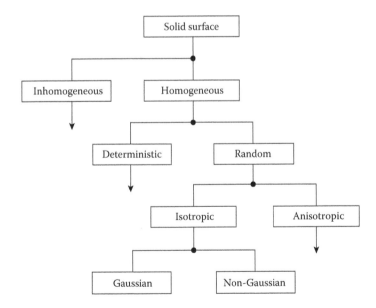

Figure 3.1 General typology of surfaces. (From Bhushan, B., Surface roughness analysis and measurement techniques, in: Bhushan, B., ed., *Modern Tribology Handbook*, CRC Press, Boca Raton, FL, 2000, 1760p.)

the surface roughness of typical sandblasted surfaces. However, the opinion of many specialists in the industry is that a rougher surface is beneficial to bond strength. Given that roughness depends directly on the surface preparation method, the investigations presented here are intended to shed new light on the subject and ultimately resolve the controversy.

According to the American National Standards Institute (ANSI/ASME B46.1-2009), the methods for measuring roughness and surface texture can be classified into three types: contact methods, taper sectioning methods, and optical (noncontact) methods. Bhushan (2001) considered only two categories: contact and non-contact methods. Taper sectioning, which is used in metallurgy and involves cutting across a surface at a low angle α to physically amplify asperity heights by a factor equal to ctg α (Sherrington et al., 1988), can be classified as an optical technique. Among the contact methods, there are stylus-type profilometers, tactile tests, measurement of kinetic friction, measurement of static friction, rolling ball measurements, and measurement of the compliance of a metal sphere with a rough surface. Optical (noncontact) methods include laser triangulation profilometry, interferometry, and optical reflecting instruments. Light microscopy and scanning electron microscopy may be counted in this group of methods.

A variety of approaches have been used over the years to characterize the surface roughness of concrete: evaluation of the proportion of the surface occupied by aggregates, measurement of the maximum roughness amplitude, adhesion tests, calculation of surface parameters based on image analysis or on microscopy observations, etc. However, many of these methods are unable to provide a sufficiently detailed representation of the actual surface profile for the calculation of morphological and statistical parameters and are not user-friendly under field conditions. In order to achieve a reliable quantitative analysis of superficial concrete morphology after surface preparation, different profilometry and surfometry techniques can be used (Fukuzawa et al., 2001; Maerz et al., 2001; Courard and Nelis, 2003a; Courard et al., 2007; Garbacz and Kostana, 2007; Santos and Júlio, 2008, 2010; Perez et al., 2009). The data obtained from such techniques makes it possible to conduct a real quantitative assessment of the surface profile by means of statistical parameters calculated from the total superficial profile and from the filtered waviness (low frequency/macroroughness) and roughness (high frequency/microroughness) profiles. Some of these parameters, for example, the arithmetic mean profile and the flatness coefficient, are particularly effective, both for the shape of valleys and peaks, as well as for their amplitude and frequency. A state-of-the-art review on roughness quantification methods for concrete surfaces was recently presented by Santos and Julio (2013). Table 3.2 presents the general characteristics of concrete surface treatments based on their consideration.

The most suitable techniques for both laboratory and field use, and the most relevant quantitative roughness characteristics, are described in the next part of this section. The following techniques are analyzed on a comparative basis, taking into account their effectiveness, accuracy, consistency, and field applicability:

- *Concrete surface profile* (CSP), according to International Concrete Repair Institute (ICRI) Guideline No. 03732
- *Sand patch test*, according to ASTM E965 (very similar to EN 13036-1:2005) and EN 1766
- *Mechanical profilometry*, in which a high-precision extensometer is moved over the entire surface to yield a 3D map (with x, y, and z coordinates) from which morphological parameters are computed
- *Laser triangulation profilometry*, in which the superficial elevation (distance from the laser beam source to the object) of each point is calculated on the basis of the laser beam transit time
- *Interferometrical profilometry*, based on observation and analysis of the shadow produced by the superficial roughness of the surface (moiré fringe pattern principle in this case)

Table 3.2 Comparison of various methods for concrete surface texture characterization

Roughness quantification method	Quantitative evaluation	Nondestructive	Cost	Portable	Work intensive	Contact with surface
Concrete surface profile	No	Yes	Low	Yes	No	No
Sand patch test	Yes	Yes	Low	Yes	No	Yes
Mechanical stylus	Yes	No	Medium	No	Yes	Yes
Laser profilometer	Yes	Yes	Medium	No	Yes	Yes
Interferometric profilometer	Yes	Yes	Medium	Yes	No	No
Shadow profilometry	Yes	Yes	Low	Yes	Yes	Yes
PDI method	Yes	No	Low	No	Yes	Yes
2D LRA method	Yes	Yes	Medium	Yes	No	No
3D laser scanning method	Yes	Yes	High	Yes	No	No
Circular track meter	Yes	Yes	Medium	Yes	No	No
Digital surface roughness meter	Yes	Yes	Medium	Yes	No	No
Microscopy	Yes	No	High	No	Yes	No
Slit-island method	Yes	No	Low	No	Yes	Yes
Roughness gradient method	Yes	No	Low	No	Yes	Yes
Photogrammetric method	Yes	Yes	Medium	Yes	Yes	No
Outflow meter	Yes	Yes	Low	Yes	No	Yes
Air leakage method	No	Yes	Low	Yes	No	Yes
Ultrasonic method	No	Yes	Medium	Yes	No	No

Source: Santos, P. and Júlio, E., *Constr. Build. Mater.*, 38(5), 912, 2013.

3.2.2 Concrete surface profile

The visual observation of surface roughness is the simplest evaluation method, but it is rather subjective. A benchmark profile method has been proposed by ICRI (ACI 562 Repair Code) to provide clear visual standards for purposes of specification, execution, and verification. The profile reference replicates that make up the CSP (Table 3.3) represent concrete surfaces after typical surface treatments commonly used in the field; details are given in the ICRI Guideline No. 03732. The range of evaluation is, however, limited to gentle surface treatments.

3.2.3 Sand patch test

The sand patch test described in ASTM E965 (similar to EN 13036-1:2005) is one of the most commonly used methods for examining the macrotexture depth of concrete surfaces, mainly for road and airfield pavements. This method consists in careful application of a specific volume and granulometry of granular materials (glass spheres or sand) onto a surface and subsequent measurement of the total area covered.

Table 3.3 Concrete surface treatment methods and corresponding CSPs (acc. to ICRI Guideline No. 03732)

Profile reference replicates	Surface preparation methods	CSP
	Detergent scrubbing	I
	Low-pressure water cleaning	I
	Acid etching	I–3
	Grinding	I–3
	Abrasive blasting (sand)	2–5
	Steel shot-blasting	3–8
	Scarifying	4–9
	Needle scaling	5–8
	Hydrodemolition	6–9
	Scabbling	7–9
	Flame blasting	8–9
	Milling/rotomilling	9

Source: ICRI Guideline nr 03732, Selecting and Specifying Concrete Surface Preparation for Sealers, Coatings, and Polymer Overlays, 2002.

Surface roughness is characterized by mean texture depth (MTD), calculated in accordance with Equation 3.1:

$$MTD = \frac{4V}{\pi D^2} \text{ (mm)}$$
(3.1)

where
  V is the volume of a granular material (mm^3)
  D is the diameter of a circle covered by a granular material (mm)

A similar method for evaluating surface roughness is proposed in the European Standard EN 1766:2000 in the case of concrete substrate preparation prior to repair. Silica sand with a granulometry of 100/50 μm is recommended for evaluation (Figure 3.2). The surface roughness index (SRI) is calculated in accordance with Equation 3.2:

$$SRI = \frac{V}{D^2} \cdot 1272$$
(3.2)

where the symbols are the same as in Equation 3.1; $V = 25$ mL is recommended.

The advantages of the sand patch method are its speed, nondestructive character, and applicability in situ; a disadvantage is that the surface has

Figure 3.2 Measurement of surface macrotexture with the sand patch test. (From Courard, L. et al., Concrete surface roughness characterization by means of opto-morphology technique, in: Czarnecki, L. and Garbacz, A., eds., *Adhesion in Interfaces of Building Materials: A Multi-Scale Approach, Advances in Materials Science and Restoration*, AMSR No. 2, Aedificatio Publishers, Freiburg, Germany, 2007, pp. 107–115.)

to be protected from wind and rain. The main limitations are the range of validity (from 0.25 to 5 mm) and the fact that it can be used only on horizontal surfaces.

3.2.4 Mechanical profilometry

In this method, deviations in surface profile are detected by a sensor (stylus) that moves along the surface (Courard and Nelis, 2003a). The gauge converts vertical deflections of the stylus position into electrical signals which are recorded by the data acquisition system, thus creating a surface profile (Figure 3.3a). It is possible to regulate the distance between measurement points for better precision. The geometry (round or conical) and size (radius) of the stylus tip are of prime significance for the profile to register; profiles characterized by small wavelengths will not be measured properly if the diamond cone radius is too large (Figure 3.3b and c).

Roughness measurements usually yield images of the profile. To analyze the influence of the treatment on the surface, it is necessary to mathematically and statistically quantify the shape of the surface by means of several parameters. Another approach is surfometry, a surface metrology of the

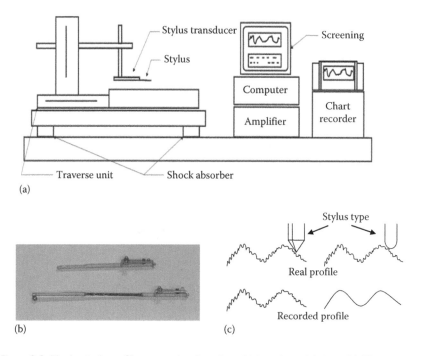

Figure 3.3 Mechanical profilometer developed at University of Liège: (a) Test setup, (b) stylus used in concrete surface roughness evaluation, and (c) schematical representation of the influence of the stylus tip geometry on the recorded profile. (Adapted from Courard, L. and Nélis, M., *Mag. Concr. Res.*, 55, 355, 2003a.)

profile rendered in 3D; in this case, the profilometer is used to obtain several profiles in parallel. The results are not analyzed in a single direction but in two directions perpendicular to each other (x, y) to generate a 3D visualization of the surface (Figure 3.4). This method yields a quantification of the surface texture independent of the anisotropy.

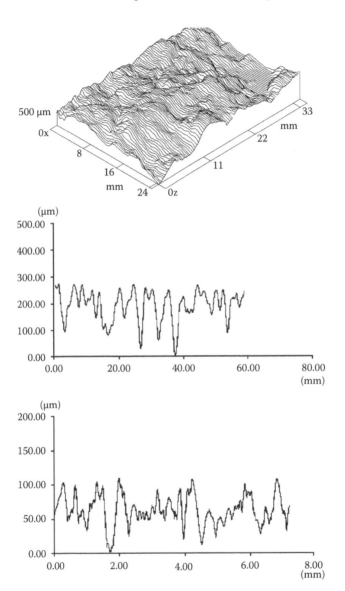

Figure 3.4 Examples of 3D visualization, waviness, and roughness profiles, respectively, after sandblasting, as determined through mechanical profilometry. (Courtesy of University of Liège, Liège, Belgium.)

3.2.5 Laser profilometry

The laser profilometry method is based on laser distance measurement by a displacement sensor (Fukuzawa et al., 2001). The most recently developed laser profilometers are fast and accurate and are able to measure surface topography down to the submicrometer level over an area of 500 × 500 mm in both 3D and 2D outputs. The technique is based on the principle of optical triangulation. Optical triangulation uses a light source (commonly a diode laser), imaging optics, and a photodetector. As shown in Figure 3.5a, a diode laser is used for generating a collimated beam of light, which is then projected onto a target surface. A lens focuses the spot of the reflected laser light onto a photodetector, which generates a signal that is proportional to the spot's position on the detector. As the target surface height changes, the image spot may shift due to the parallax. The sensor scans in two dimensions to generate a 3D image of the part's surface (Figure 3.5b).

The first applications of commercial laser profilometry were used to characterize surface texture for tribology. The technique has also been used to characterize concrete surfaces (Fukuzawa et al., 2001; Garbacz and Kostana, 2007; Santos and Júlio, 2008).

The recently developed circular track meter (CTM), in which a CCD laser displacement sensor is used, belongs to this group of profilometers. The eight individual segments are analyzed to investigate the profile at

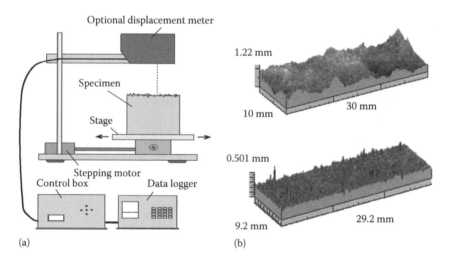

(a)   (b)

Figure 3.5 Laser profilometry with an optical displacement meter: (a) principle of measurement (From Fukuzawa, K. et al., Surface roughness indexes for evaluation of bond strengths between CRFP sheet and concrete, in: Fowler, D., ed., *10th International Congress on Polymers in Concrete (ICPIC'01)*, Honolulu, Hawaii, Paper No. 12, 2001.) and (b) 3D visualization of waviness and roughness of a sandblasted concrete substrate. (Courtesy of Warsaw University of Technology, Warsaw, Poland.)

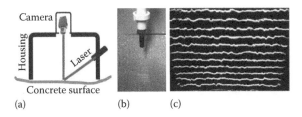

Figure 3.6 Laser profilometry—schematic representation of the laser profiling equipment (a), line laser (b), and image of a concrete surface (c). (From Maerz, H. et al., Concrete roughness characterization using laser profilometry for fiber-reinforced polymer sheet application, *80th Annual Meeting*, Transportation Research Board, Washington, DC, Paper No. 01-0139, 2001.)

angles that are parallel, perpendicular, and 45° to the direction of travel. The CCD is mounted on an arm which is driven by a DC motor and rotates at 80 mm above the surface with a 142 mm radius. The data are segmented into eight 111.5 mm arcs of 128 samples each. The results of profile characterization with CTM are presented in the form of mean profile depth and the root mean square. The details are given in the ASTM E2157-01 standard.

Optical sectioning method, also called the Schmaltz technique, is a similar optical technique. Maerz et al. (2001) developed a portable concrete roughness testing device consisting of an optical laser-based imaging system that operates in accordance with the principles of the Schmaltz microscope and the shadow profilometry method. It uses a laser profiling line (laser striping) that produces a non-Gaussian (i.e., uniform) distribution of light intensity along the line. The investigated concrete surface is illuminated with thin slits of red laser light at an angle of 45°, while the observations are performed perpendicularly to the surface (Figure 3.6). A high-resolution (tiny) board CCD camera with a 7.5 mm lens is fixed vertically on the protection housing.

3.2.6 Interferometrical profilometry

Recent studies have been devoted to optical methods (Figure 3.7a) that can be used reliably in field applications (Courard et al., 2007; Siewczyńska, 2008). Among other optical interferometry methods, the moiré projection technique can be exploited advantageously for this purpose. The moiré phenomenon appears when two networks of light rays made of equidistant lines (alternatively opaque and transparent) are superimposed. The technique of relief identification is based on the measurement of a parallel fringe pattern from a deformed pattern projected on a non-plane surface (Figure 3.7b). The moiré fringes are similar to level lines representing the height variations of the object. When a network of parallel fringes is projected onto a plane

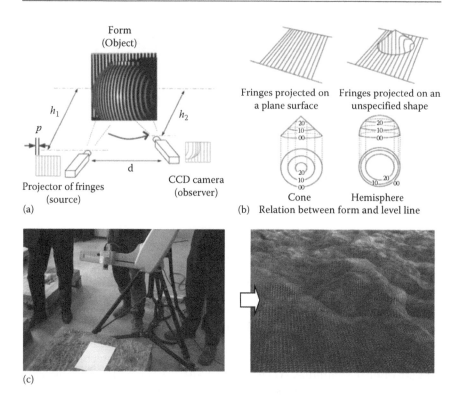

Figure 3.7 Optomorphological profilometry—principle of measurement (a), relationship between form and level line (b), and testing system and example of 3D visualization of concrete surface (c). (Courtesy of University of Liège, Liège, Belgium.)

surface, it will not be deformed, but when projected onto an unspecified shape, this same network will be deformed. The main principle of the test is to compare two images with different moiré networks. The first image is the reference; it corresponds to the network of non-deformed parallel fringes. The second image contains the projected network deformed according to the non-plane shape. An algorithm analyzes the image and compares the grid of calibration and the deformed grid.

3.2.7 Profile description

After treatment, concrete surfaces are characterized by fractal topography. As with any fractal object, it is possible to break up this surface or this profile into a series of subprofiles. Each subprofile can be differentiated in terms of wavelength; there are, however, no limits or precise criteria involved in validating the choice of decomposition method (Figure 3.8). It is also possible to filter the results mathematically, as shown by Perez et al. (2009).

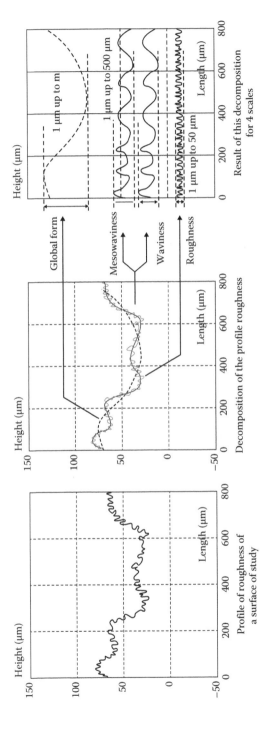

Figure 3.8 Effect of scales on profile decomposition. (From Perez, F. et al., *Mag. Concr. Res.*, 61(6), 389, 2009.)

Table 3.4 Two components of total surface profile

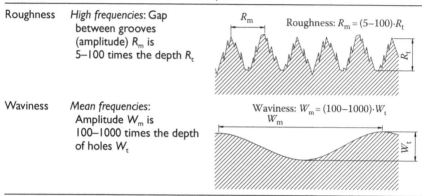

Roughness	*High frequencies:* Gap between grooves (amplitude) R_m is 5–100 times the depth R_t
Waviness	*Mean frequencies:* Amplitude W_m is 100–1000 times the depth of holes W_t

Roughness: $R_m = (5–100)\cdot R_t$

Waviness: $W_m = (100–1000)\cdot W_t$

Source: Courard, L. and Nélis, M., *Mag. Concr. Res.*, 55, 355, 2003a.

In their study, since the two surfometry methods (mechanical and interferometrical) had different resolution levels, it was possible to obtain complementary scales of topography. The method with a mechanical stylus and high resolution yields roughness (R) and waviness (W) (Table 3.4) (Courard et al., 2003). With the interferometrical method at a resolution of 0.200 μm, it is possible to obtain two higher scales named meso-waviness (M) and shape (F). In mechanical profilometry, a filtration by differentiating the diameters of the stylus is often used.

According to EN ISO 4287, the total (primary) profile, the waviness and roughness profiles, can be characterized by several vertical (Table 3.5) and horizontal (Table 3.6) amplitude parameters. Surface parameters are determined on the basis of the mean line as a reference line; this reference is usually defined in such a way that, in the limits of the profile length, the sum of the squared values of the altitudes of the profile measured versus this reference line is minimal.

Using horizontal profile parameters, an Abbott curve, also called a bearing curve (Courard and Nélis, 2003), can be determined. This provides information about the surface profile: a gradual decrease in the curve suggests a surface with few holes, while a more steeply decreasing curve is characteristic of a surface with a lot of holes. Important parameters for analyzing the distribution of holes and peaks, as well as the shape of the profile can be graphically calculated from the Abbott curve (Table 3.7). These parameters are crucial when it comes to evaluating the quantity of slurry, mortar, etc. needed for the interface area between the concrete substrate and the new layer (Figure 3.9).

Concrete surface texture can be characterized using a scientific approach called quantitative fractography, which is based on image analysis of cross-sectional profile (Underwood, 1987; Gokhale and Drury, 1990). This approach is well developed in the case of metals and ceramics in comparison with concrete-like composites (Wojnar, 1995; Krzydłowski and

Table 3.5 Vertical amplitude parameters of surface profile as per EN ISO 4287

Symbol	Parameters	Definition
m_x	Mean value and line	Line whose height (mean value) is determined by minimal sum square deviation of the profile defined as follows: $X = \min \sum y^2(x)$
X_p	Maximum peak height	Distance between the highest point of the profile and the mean line
X_m	Minimum valley depth	Distance between the lowest point of the profile and the mean line
X_t	Maximum height	Maximum distance between the lowest and the highest point of the profile and its equal: $X_t = \max(X_p + X_m)$
X_a	Arithmetic mean deviation	Mean departure of the profile from the reference mean line as follows: $X_a = \frac{1}{l} \int_0^l |y(x)|\, dx$, approximated by $X_a \approx \frac{1}{n} \sum_{i=1}^n |y_i|$
X_q	Root mean square deviation	Statistical nature parameter defined in the limits of the cut-off length as follows: $X_q = \frac{1}{l} \int_0^l y^2(x)\,dx$

Source: EN ISO 4287, Geometrical product specifications (GPS)—Surface texture: Profile method—Terms, definitions and surface texture parameters, 2009.

R, roughness; W, waviness; P, total profile (instead of index "X").

Ralph, 1996). However, geometrical and stereological parameters are also of significant importance in concrete-like composites (Saouma and Barton, 1994; Stroeven, 2000; Czarnecki et al., 2001; Siewczyńska, 2008). Besides the profile parameters determined as per EN ISO 4287, two additional ste-reological parameters could be considered for characterization of concrete

Table 3.6 Horizontal amplitude parameters of surface profile as per EN ISO 4287

Symbol	Parameters	Definition
S_k	Skewness of surface height distribution	A measure of asymmetry of profile deviations about the mean line, as follows: $S_k = \dfrac{1}{R_q^3} \dfrac{1}{n} \sum_{i=1}^{n} Y_i^3$
S_m	Mean period of profile roughness	Mean value of mean line consecutively including a peak and a valley S_{m}, as follows: $S_m = \dfrac{1}{n} \sum_{i=1}^{n} S_{mi}$
n_p	Bearing length	Sum of partial lengths n_i corresponding to the profile cut by a line parallel to the mean one for a given cutting level
t_p	Bearing length ratio	Ratio between bearing length and cut-off length, expressed as a percentage: $t_p = n_p/l$

Source: EN ISO 4287, Geometrical product specifications (GPS)—Surface texture: Profile method—Terms, definitions and surface texture parameters, 2009.

Table 3.7 Abbott curve parameters

Symbol	Parameters	Definition
C_R	Relative height of the peaks	Gives an idea of significance of the volume of very high peaks above the reference line
C_F	Depth of the profile	Excluding high peaks and deep holes gives information on surface flatness; a lower value of C_F indicates great surface flatness
C_L	Relative depth of the holes	Gives an idea of the significance of the volume of voids under the reference line

surfaces after surface treatment (Kurzydłowski and Ralph, 1996; Garbacz et al., 2006):

- Profile (linear) roughness ratio R_L: Length of the profile line L divided by the projected length of the profile line L_O:

$$R_L = \frac{L}{L_O} \qquad (3.3)$$

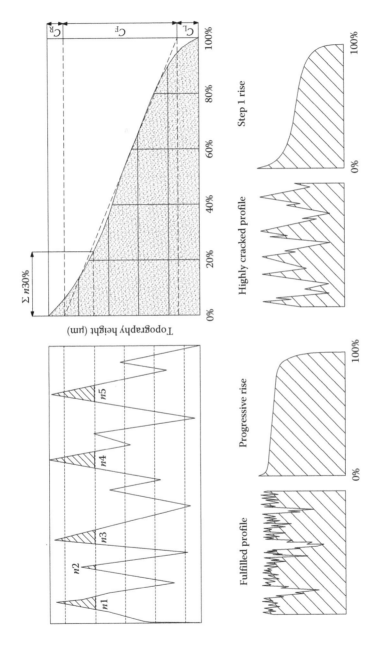

Figure 3.9 The Abbott curve and its interpretation.

- Surface roughness ratio R_S: True fracture surface area S divided by the apparent projected area S_O:

$$R_S = \frac{S}{S_O} \tag{3.4}$$

- Fractal dimension D as a measure of this self-similarity of the profile line. The basic requirement for the fractal object is that some structural feature or unit is sequentially repeated at different levels. The geometry of the fracture surface of concrete-like composites is related to the scale of the observation. This implies that the self-similarity of the fracture surface may not be extended over all ranges of magnification.

These parameters can be determined from the image of the profile reproduced from the sample cross-section or the profile recorded with profilometry.

3.2.8 Relationship between parameters determined with different techniques

Several concrete substrates with different compressive strengths and textures obtained after surface preparation were tested at Laval University, the University of Liège, and the Warsaw University of Technology.

3.2.8.1 Mechanical versus laser profilometry

A comparative study of surface roughness characterization with laser and mechanical lab profilometers was conducted on low-strength concrete substrates of C20/25 subjected to a range of surface preparation methods with various levels of agressiveness (Garbacz et al., 2005). As a result, significantly different roughness of surfaces was obtained. The total profile obtained was then filtered and decomposed into low and high frequencies to separate parameters of waviness and roughness, respectively. Further in the text, indexes p and s denote parameters measured by mechanical and laser profilometer, respectively.

The statistical analysis of the results revealed a high correlation coefficient ($r > 0.94$) of the relationship between the corresponding mean values of the waviness profile, W_a (Figure 3.10a) as well as the Abbott's parameters C_R and C_F determined with laser and mechanical profilometry (Figure 3.10b). A higher scatter in the results for both profilometry methods is observed in the case of other amplitude parameters. Lower statistical significance (Figure 3.10c) is obtained for the total heights of the waviness profile (W_{ts} vs. W_{tp}) and the maximum depth of the valleys (W_{vs} vs. W_{vp}) as well as the relative depth of the holes, C_L (see Figure 3.10b). This could be caused by the fact that different surface areas were scanned with the laser and the mechanical profilometer. However, Figures 3.10b and c indicate that the

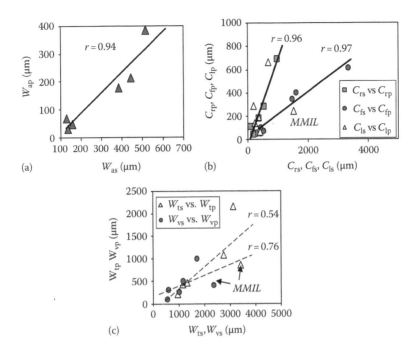

Figure 3.10 Relationships between waviness parameters: W_a (a), Abbott's (b), and W_t and W_v (c) determined with laser and mechanical profilometry; suffixes "p" and "s" for mechanical and laser profilometers, respectively. (From Garbacz, A. et al., *Mater. Charact.*, 56, 281, 2006.)

low correlation is due to the low values of amplitude parameters obtained with mechanical profilometry for the surface after mechanical milling. This surface has high irregularities and a significant number of deep and wide cracks. It seems that these cracks might be more easily detected by the laser profilometer than by the mechanical profilometer stylus.

3.2.8.2 Profilometry versus microscopic method

The relationship between R_S and W_{as} and W_{ap} had a relatively low correlation coefficient—r close to 0.8 (Figure 3.11). This can be explained by the fact that the stereological parameter R_S was calculated for a longer profile length compared to the profile length of the sample tested with laser profilometry (350 vs. 50 mm).

3.2.8.3 CSP profiles versus optical measurements

The surface geometry characteristics obtained with optical profilometry and the visual method (CSP profiles) were compared. A profile obtained through this approach gave the description of meso-waviness and overall shape. Figure 3.12

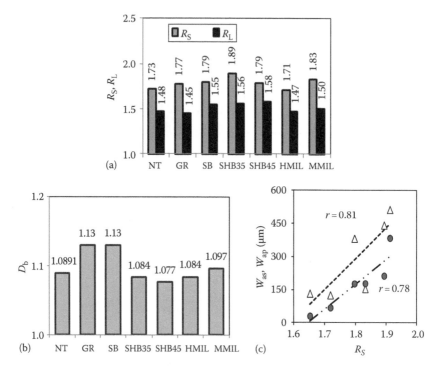

Figure 3.11 Stereological parameters of tested concrete substrates: (a) R_S and R_L, (b) D_b for concrete substrates after various surface treatments, and (c) the relationships R_S vs. arithmetic mean deviation of waviness profile determined with laser (W_{as}) and mechanical (W_{ap}) profilometers. (From Garbacz, A. et al., *Bull. Pol. Acad. Sci. Tech. Sci.,* 61(1), 73, 2013.)

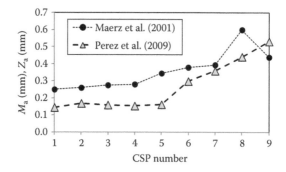

Figure 3.12 Quantitative roughness evaluation of ICRI's CSP plaques: arithmetic mean deviation M_a calculated by Perez et al. (2009) and root mean square of the first derivative of the profile Z_a calculated by Maerz et al., (2001). (From Garbacz, A. et al., *Bull. Pol. Acad. Sci. Tech. Sci.,* 61(1), 73, 2013.)

shows that the optometric device is not able to detect any change in terms of roughness level under a threshold CSP value (no. 5) corresponding to the vertical resolution of the optometric device. Nevertheless, above this value, the optometric method accurately reproduces the surface roughness level in accordance with the CSP scale. Similar investigations were performed by Maerz et al. (2001).

It can be concluded that it is possible to significantly improve the CSP replicate system through a real quantitative approach. The actual CSP plates are rather narrow with respect to the spectrum of CSPs obtained with actual surface preparation techniques. The identification of reference curves, similar to those presented by Perez et al. but on a wider scale of surface roughness, will help broaden the range of application of this method to much coarser profiles such as those obtained with jack-hammering and water-jetting, for example.

3.2.8.4 SRI versus profilometry parameters

Figure 3.13a and b presents a comparison between SRI (*Surface Roughness Index*) values and parameters determined using more sophisticated profilometry techniques: an equivalent correlation exists between the mean

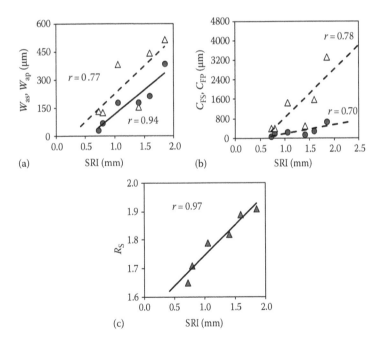

Figure 3.13 Surface Roughness Index vs.: (a) arithmetic mean of waviness, (b) Abbott's parameters, and (c) R_S ratio; (p, Δ) and (s, •) for mechanical and laser profilometers. (From Garbacz, A. et al., *Bull. Pol. Acad. Sci. Tech. Sci.*, 61(1), 73, 2013.)

waviness obtained by means of the two profilometry techniques and SRI, respectively. Similar conclusions may be given for Abbott's curve parameters (Figure 3.13b). The relationship between Rs and SRI exhibits a very high correlation coefficient $r = 0.97$ (Figure 3.13c). This confirms that SRI is a good estimation of the mean deviation of a concrete surface profile after preparation and that it can be used for practical evaluation of surface roughness.

3.2.9 Summary

Recently, various techniques have become available for the characterization of concrete surface texture. The combination of different methods enables a very good description of "roughness" at various scales. Depending on what has to be analyzed, mechanical and laser profilometers are more accurate for microroughness, while the optical method seems to give a better description of the shape of the profile. However, investigations with very precise laboratory laser and mechanical profilometers of concrete surfaces after various preparations in terms of their aggressiveness clearly indicated that the surface preparation technique has no major influence on microroughness ("high frequencies waves"). This allows concluding that waviness parameters are enough for the assessment of concrete surface prior to repair. This shows the usefulness of the recently developed optical method, for example, based on the moiré pattern. The main advantages are the rapidity of the procedure and the large area that can be observed in one operation.

3.3 MECHANICAL PROPERTIES

The evaluation of the strength of superficial concrete is always needed (Naderi et al., 1986). It can be based on both destructive and non-destructive testing, which are often used complementarily:

a. Compressive strength determination on cores drilled from the member or structure.
b. *Rebound hammer*: The test hammer rebound after impact with the surface is dependent on the concrete hardness and can thus be correlated to compressive strength. It is recognized as a useful tool for performing quick surveys to assess the uniformity of concrete. However, because of the many factors, besides concrete strength, that can affect the rebound number, it is not generally recommended where accurate strength estimates are needed (Courard et al., 2012).
c. *Ultrasonic pulse velocity (UP-V)*: The speed of compressional waves in a solid is related to the elastic constants (modulus of elasticity and Poisson's ratio) and the density. By conducting tests in various areas

of a structure, locations with lower quality concrete can be detected by lower pulse velocity (Carino, 2003).

d. *Probe penetration*: The probe penetration method involves using a gun to drive a hardened steel rod, or probe, into the concrete and measuring the exposed length of the probe, which can be used to estimate compressive strength.

e. *Pullout test*: The force required to pull out an insert is an indicator of concrete strength.

f. *Break-off test*: This test measures the force required to break off a cylindrical core from the concrete mass.

g. *Pull-off test*: The results of this test not only provides an evaluation of the cohesion of the superficial layer of concrete but also an estimation of the concrete integrity (Cleland et al., 1986). The presence of cracks parallel to the surface, induced or not by surface preparation (Courard et al., 2004), can be easily detected.

Main advantages and disadvantages of several existing NDT are presented in Table 3.8.

3.3.1 Compressive strength

Compressive strength may be evaluated directly on cores collected by drilling or sawing (EN 12504-1). At test performed in a laboratory, according to EN 12390-3, ASTM C109, or ASTM C873, allows having a more accurate evaluation of compressive strength. One must take into account the dimensions of the samples with regard to the maximum aggregate size and the ratio between the diameter and the height of the sample; for drilled cores, the preferable condition is that the core diameter is at least three times the maximum aggregate size (Bungey, 1989). Moreover, the accuracy of the compressive test results has been found to decrease as the diameter-to-aggregate ratio decreases (Malhotra, 1977). A length-to-diameter ratio of between 1.5 and 2.5 is recommended for compressive strength evaluation (STP 169C, 1994).

3.3.2 Schmidt rebound hammer

This test, performed in accordance with ASTM C805 or EN 12504-2, is particularly well adapted for near-to-surface layer investigations as it concerns only the upper surface of the concrete. Figure 3.14 presents the average results and coefficients of variation (C.V.) (Courard et al., 2012), respectively, of the Schmidt hammer soundings performed on different testing surfaces (average of 60 results for reference and sandblasting treatment (SB); average of 25 results for treatments with a concrete breaker (CB)).

Compressive strength values calculated from the recorded Schmidt hammer rebound data are strictly used here on a comparative basis. As shown in

Table 3.8 Principles and advantages of semi- and nondestructive investigation techniques

Methods and scheme	Principle	Application	Limitations
Acoustic impact: Various equipment from hammers or drag chains to electronic systems with auditory system	Impact of the object surface with an implement; frequency and dumping of sounds indicate the presence of defects.	Detection debonds, delaminations, voids	• Geometry and mass of object influence the results. • Low discrimination for auditory system.
Break-off:	Measurements of the force required to break off a cylindrical core from the concrete mass.	• Evaluation of compressive strength in both new and existing construction • Estimation of bond strength between concrete and overlays	• Not recommended for concrete with max. aggregate size greater than 25 mm. • Difficulties (sleeve insertion) during evaluation of new structure.
Electrical/magnetic methods (covermeters):	Interactions between the bars and low-frequency electromagnetic fields. Two principles: magnetic reluctance and eddy currents.	• Locate reinforcing bars and estimate the diameter and depth of the cover • Evaluation of moisture content in the concrete	• Expensive and specialized equipment. • Results depend on salt content and temperature changes.

(Continued)

Table 3.8 (Continued) Principles and advantages of semi- and nondestructive investigation techniques

Methods and scheme	Principle	Application	Limitations
Fiber optics (video-endoscopy)	Using of a fiber-optic probe consisting of flexible optical fibers, lens, and illuminating system is inserted into drilled hole, cracks, new generation consisting of additional CCD chip to improve image.	Observation of inaccessible areas or within elements of the structure for the detection of defects (cracks, voids, corrosion, etc.), possibility of multi-directional observation, high-resolution images, possibility to record with camera	• Expensive equipment. • Necessity for many holes to have adequate access.
Infrared thermography (IT):	Measurement of surface temperature differences—thermographic image.	Locating near-surface defects, like voids delamination and flaws	Results depend on in situ conditions (surface quality, wind speed, and ambient temperature).

Infrared scanner

(Continued)

Table 3.8 (Continued) Principles and advantages of semi- and nondestructive investigation techniques

Methods and scheme	Principle	Application	Limitations
Impact-Echo, I-E: $d = C_p / 2f_d$	Development of echo methods; mechanical, high energy impact used to generate the stress wave; high penetration of concrete, mainly by P wave; frequency analysis of recorded waveform using the fast Fourier transform.	• Evaluation of thickness of concrete slab and thick overlays • Defect detection in concrete slabs, like delamination, flaws, large cracks, honeycombing, debonding, including quality of the bond between overlay and base concrete • Estimation of the depth of surface-opening cracks, including those filled with water • Detection and localization of other elements of the floor system like waterproofing isolation	• Detection of large defects, relatively deeply located. • Needs an expert for interpretation of results.
Impulse-response method, I-R: 	A low-strain impact with an instrumented, rubber-tipped hammer produces a stress wave. Both the time trace of the hammer force and the velocity transducer are processed into frequency using the fast Fourier transform. The "mobility" plot is constructed.	• Methods developed for evaluation of deep foundations and mass concrete • Stiffness measurements • Quick scanning for flaws detection of concrete structures before lateral detail analysis, e.g., with I-E method	• Results interpretation is delicate. • Unable to determine defects in shafts > 30 m or with L/d > 30.

(Continued)

Table 3.8 (Continued) Principles and advantages of semi- and nondestructive investigation techniques

Methods and scheme	Principle	Application	Limitations
Ground-Penetrating Radar (GPR):	Noncontact method; method analogues to UP-E techniques, except that pulses of electromagnetic waves are used instead of stress waves, results are recorded as a water plot; can be used when only one side is available.	• Detection of delamination • Locating reinforcing bars in structures • Measurement of pavement • Thickness • Measurements of water content of fresh concrete	• Improper estimation of relative dielectric constant resulted in a large error. • Needs an expert for interpretation of results. • Results depend on in situ conditions, e.g., presence of moisture and chlorides in concrete. • Expensive equipment.
Nuclear (radioactive) methods:	Methods involve a source of penetration electromagnetic radiation and a sensor to measure the intensity of the radiation after it has travelled through the object. The two techniques available for industrial floors: direct transmission and backscatter radiometry.	• Evaluation of density of fresh and hardened concrete • Could be used for reinforcement detection	• Operator must be licensed. • Available equipment limited to path less than 300 mm (direct transmission) and 100 mm (backscatter r.). • Direct transmission need drilling a hole and inserting the source of radiation.

(Continued)

Table 3.8 (Continued) Principles and advantages of semi- and nondestructive investigation techniques

Methods and scheme	Principle	Application	Limitations
Probe penetration:	The method involves using a gun to drive a hardened steel rod, or probe, into the concrete; the exposed length of the probe is used to estimate compressive strength.	Mainly for estimation of compressive strength	• Results depend on hardness of aggregate. • Reinforcing steel affects results; location of reinforcement bars should be determined before test.
Pull-off:	Measurement of the force required to pull off a glued steel disk from the drilled layer or layers; this force is a measure of bond strength in multilayer systems or tensile concrete strength.	• Evaluation of bond strength in multilayer systems • Evaluation of tensile strength of concrete substrate • Useful for both lab and onsite investigations	• Needs minor repair in the test place. • Results sensitive on drilling depth and inclination from perpendicular direction.
Pullout:	Measurement of the force required to pull out a steel anchor from the concrete substrate; this force is a measure of concrete strength.	• Estimation of concrete strength in both new and existing structures • Monitoring of concrete strength gain	• Special equipment is necessary to install the anchor in the concrete. • Needs minor repair in the test place.

(Continued)

Table 3.8 (Continued) Principles and advantages of semi- and nondestructive investigation techniques

Methods and scheme	Principle	Application	Limitations
Rebound hammer  $f_c = f(L)$; L, rebound index	Measurement of rebound height after striking the concrete surface with spring loaded hammer; the correlation between rebound index and compressive strength is determined.	• Evaluation of concrete substrate homogeneity • Estimation of concrete substrate compressive strength according to common procedure for concrete structure • Monitoring strength gain	• Evaluation of near-surface properties only. • Results depend on surface roughness. • Reference curve needed for strength estimation. • Rebound number affected by the orientation of the apparatus.
Spectral analysis of surface waves (SASW): R Imp act R wave wave X	Analysis of the spectrum of the disperse generalized Rayleigh surface wave in a layered system; the received signal is analyzed to obtained the dependence of phase velocity on the frequency.	• Determination of the stiffness profiles of flexible pavements • Measurement of elastic properties of layered systems, such a pavements • Interlayered good or poor concrete	• Necessity for comparison of the theoretical and experimental dispersion curves. • Time-consuming procedure. • Difficulties in interpretation of the results.

$\lambda_f = 360X/\varphi_f$, λ_f—wavelength for component
frequency f, X—distance between receiver;
φ_f—phase angle of component with frequency f

(*Continued*)

Table 3.8 (Continued) Principles and advantages of semi- and nondestructive investigation techniques

Methods and scheme	Principle	Application	Limitations
Ultrasonic pulse echo (UP-E): $d = v_p t/2$	Propagation of a short pulse of ultrasonic wave; measurement of travel time to boundaries separating materials with different densities and elastic properties; by knowing the wave speed, v_p, the distance to the reflecting interface is calculated.	• Method developed to detect delamination, discontinuities, and small cracks • Measurements of slab thickness • Monitoring of polymer adhesive curing • Evaluation of thickness of polymer coating	• Needs coupling agent. • Heterogeneous nature of concrete and of reinforcement presence result in multitude of echoes. • Difficult interpretation of results. • Relatively large "dead-zone".
Ultrasonic pulse velocity, UPV Direct Indirect 	Measurement of the travel time of ultrasonic P wave, over a known path length, calculation of pulse velocity in concrete, analysis of the relationship between pulse velocity and concrete properties (mainly compressive strength).	• Evaluation of concrete substrate homogeneity • Estimation of concrete substrate compressive strength • No commonly accepted procedure for the estimation of crack depth and determination of dynamic Young's modulus	• Needs a coupling agent. • Reference curve needed for strength estimation. • Measurements with indirect method difficult to interpret.

Source: Courard, L. et al., Evaluation and quality assessment of industrial floors (Chapter 4), RILEM TC 184—IFE Industrial floors for withstanding environmental attacks, including repair and maintenance. RILEM Report 33 (Ed. P. Seidler), RILEM Publication, 2006b, pp. 59–89.

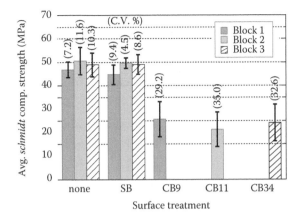

Figure 3.14 Average compressive strength values estimated from the Schmidt rebound hammer tests on slab specimens after different surface treatments (SB: sandblasting; CB9: 9 kg concrete breaker; CB11: 11 kg concrete breaker; CB34: 34 kg concrete breaker).

Figure 3.1, the results obtained for the surfaces prepared with concrete breakers exhibit much more variability, which can be attributed to the following:

- Variability in the procedure (applied force and duration)
- Angle between the axis of the hammer and the concrete surface
- Surface composition (the hammer tip can hit an aggregate, cement paste, or both)

Although this test can yield significant average values when performed over large surfaces, the data recorded (Courard et al., 2012) suggest that variability, not in as much as the absolute values, provide a reliable indication of the presence and importance of defects in the substrate. Based upon the results generated with the various investigated surface preparation methods, it appears that a threshold C.V. value of the order of 15%–20% could discriminate between prepared surfaces where significant bruising may or may not be left.

3.3.3 Ultrasonic pulse velocity

The ultrasonic pulse velocity (UP-V) method (EN12504-4 or ASTM C597) is a stress wave propagation method (see Table 3.8). Ultrasonic pulses are introduced into the concrete by a piezoelectric transducer, and a similar transducer acts as a receiver to monitor the surface vibration caused by the arrival of the pulse (Figure 3.15). A timing circuit is used to measure the time it takes for the pulse to travel from the transmitting to the receiving transducers. The speed of compressional waves

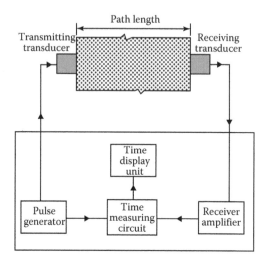

Figure 3.15 Schematic of ultrasonic pulse velocity method.

in a solid is related to the elastic constants (modulus of elasticity and Poisson's ratio) and the density.

By conducting tests in various areas of a structure, locations with lower quality concrete can be detected by their lower pulse velocity.

In summary, the UP-V method is a relatively simple test to perform on site, provided it is possible to gain access to both sides of the member. While tests can be performed with the transducers placed on the same surface, the results are not easy to interpret and this measurement generally is not recommended (Carino, 2003). Care must be exercised to assure that good and consistent coupling is achieved with the concrete surfaces. Moisture content, presence of reinforcement parallel to the pulse propagation direction, and presence of cracks or voids can affect the measured UP-V. When making UP-V measurements, the following are recommended (Courard et al., 2011a):

- To avoid cast surfaces
- To make a careful visual examination of compressive zones in the unit under test
- Where possible, to make direct transmission measurements
- To avoid heavily reinforced zones
- To make measurements at least 100 mm from the edges
- To make at least three measurements

Because of these factors, the UP-V should be used for estimating concrete strength only by experienced individuals. Like the rebound hammer test, the pulse velocity method is very useful for assessing the uniformity of concrete in a structure. It is often used to locate those portions of a structure where other tests should be performed or where cores should be drilled.

3.3.4 Probe penetration

The probe penetration method (ASTM C803) involves using a gun to drive a hardened steel rod, or probe, into the concrete and measuring the exposed length of the probe. In principle, as the strength of the concrete increases, the exposed probe length also increases, which means that the exposed length can be used to estimate compressive strength.

Probe penetration is not strongly affected by near-surface conditions and is before not as sensitive as the rebound number method to surface conditions. The direction of penetration is not important, provided that the probe is fired perpendicular to the surface. Care must be exercised when testing reinforced concrete to assure that the test is not carried out in the vicinity of the reinforcing steel, especially if the concrete cover is low (Carino, 2003).

3.3.5 Pullout test

The test can be performed according to ASTM C900 or EN 12504-3. The version that is interesting here is the post-installed version (so called Capo test). In this method, an insert with an enlarged head is sealed in the concrete. The insert and the accompanying conical fragment of concrete are extracted by using a tension-loading device reacting against a bearing ring that is concentric with the insert. The force required to pull out the insert is an indicator of concrete strength. Various procedures have been studied to achieve the anchorage of the insert (Carino, 2003).

3.3.6 Break-off test

This test (ASTM C1150) measures the force required to break off a cylindrical core from the concrete mass (see Table 3.8). The test subjects the concrete to a slowly applied force and measures a static strength of concrete. The correlations between break-off strength and compressive strength (Figure 3.16) have been found to be non-linear (Barker and Ramirez, 1988), which is in accordance with the usual practice of relating the modulus of rupture of concrete to the square root of compressive strength. Due to the relatively small size of the core and the heterogeneous nature of the concrete, the coefficient of variation of the results is about 9% (Carino, 2003).

3.3.7 Pull-off test

The pull-off test (Figure 3.17) is useful to estimate adhesion of repair systems on a concrete structure (Austin and Robins, 1993; Cleland and Long, 1997) on site or in the laboratory. This test can be performed according to ASTM D4541, ASTM D 7234, or EN 1542).

An experimental program has been conducted (Courard et al., 2004) to evaluate the influence of various test parameters on the measured cohesion of a reference concrete surface: transfer plate thickness and diameter, core

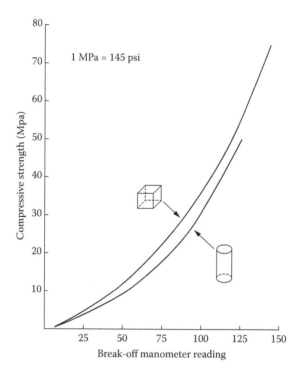

Figure 3.16 Manufacturer's correlation curves. (From Barker, M.G. and Ramirez, J.A., *ACI Mater. J.*, 85(4), 221, 1988.)

Figure 3.17 On site pull-off test device (www.safeenvironments.com.au).

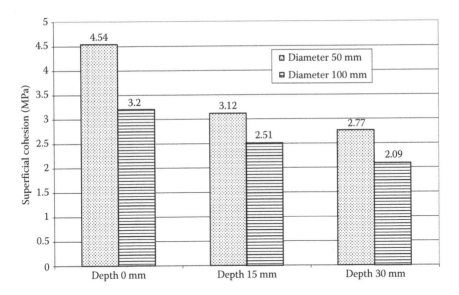

Figure 3.18 Effects of coring depth and diameter on the value of superficial cohesion of concrete. (From Courard, L. and Bissonnette, B., *Mater. Struct.*, 37(269), 342, 2004.)

drilling depth, speed of loading, adhesive type and thickness, and number of tests. A multivariate statistical analysis of the test results has shown that plate diameter and core depth are the most significant parameters, presumably with threshold values, and that there exists a synergetic effect between them (Figure 3.18). The effect of misalignment has been proved to be of low influence if the deviation angle is lower than 4°, which can be detected by the human eye (Courard et al., 2014).

The results of this test will not only provide an evaluation of the cohesion of the superficial layer of concrete but also an estimation of the concrete integrity (Bungey and Madandoust, 1992). Presence of cracks parallel to the surface, induced or not by surface preparation (Bissonnette and Courard, 2004; Courard et al., 2014), will be easily detected.

3.3.8 Test combination

Surface preparation is often a critical step in concrete repairs. While it is well acknowledged that the concrete removal operation can induce bruising and cracking in the substrate, there are still no simple practical means available to assess the integrity of a prepared surface.

Although the Schmidt rebound hammer test cannot systematically yield a reliable evaluation of the in-place compressive strength of concrete, it was shown to provide valuable comparative data for detecting superficial defects on a concrete surface (Courard et al., 2011). The rebound hammer method is thus recognized as a useful tool for performing quick surveys to assess concrete uniformity (Courard et al., 2012) and mechanical integrity over freshly prepared substrates.

The Capo pullout test has limited interest for surface preparation as it can only be carried out on smooth surfaces. Conversely, the accelerated cohesion test (Courard et al., 2012) exhibited interesting potential as a simple tool for assessing the mechanical integrity of a concrete surface prior to repair. Not only can it be used on any concrete surface, but it is also simpler and much faster than the pull-off test. Obviously, the test procedure requires some optimization; within the variables investigated in this study, the most reproducible results were obtained with a steel-threaded rod having a diameter of 9.5 mm and anchored in a 15 mm deep drilled hole. In the quest for such a test for the field evaluation of surface concrete integrity, the use of commercially available chemical anchors would certainly be desirable.

The pull-off test provides results that are very close to the actual splitting-tensile strength of the material. Moreover, it was shown in a previous study (Bissonnette and Courard, 2004) that it can effectively capture the presence of bruising. Still, it is difficult to perform adequately on vertical or overhead surfaces (Courard et al., in press), and in practice, its use is essentially limited to horizontal surfaces.

Finally, it appears that the combination Schmidt hammer/pull-off tests can fulfill the needs for the evaluation of horizontal surfaces after concrete removal, whereas the combination Schmidt hammer/accelerated cohesion tests can be used effectively on any surface, irrespective to its inclination. For quality control purposes, acceptance criteria could be specified for both the hammer soundings (ex. C.V. < 20%) and cohesion strength test results (ex. pull-off test: cohesion strength > $0.75f_{st}$; accelerated cohesion test: C.V. < 20%).

3.4 CRACKING

The cracks must be evaluated with regards to their length and width but also to their potential activity. Crack activity measurements needed for determining, in more detail, the degree of slab rocking and the load transfer characteristics across the crack can be achieved by means of extensometers and gauges that will follow vertical and/or horizontal movements (Figure 3.19).

NDT methods can also be used for cracking detection in the near-to-surface layer: UPV, impact-echo, infrared, radar are mainly applied for flaw and void detection; depending on the wavelength, cracks can also be identified and zones with internal or subsurface cracking can be identified (Van der Wielen et al., 2012).

These should be recorded and compared to the acceptable values, regarding global structural behavior of the structure and local deformations of concrete elements. Mapping of cracks can be evaluated by means of microscopical observations on cores (Bissonnette et al., 2006). Microscopic observations can be made using 100 mm (4 in.) diameter cores (about 150 mm

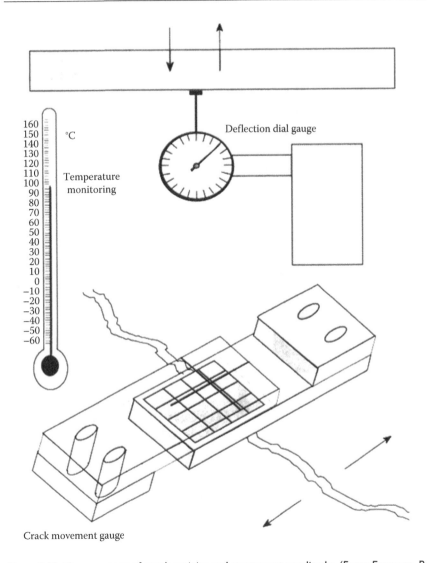

Figure 3.19 Measurement of crack activity and movement amplitude. (From Emmons, P. and Emmons, B., *Concrete Repair and Maintenance Illustrated: Problem Analysis, Repair Strategy, Techniques*, RSMeans, Kingstone MA, 1993, 295p.)

[6 in.] long) taken from the concrete blocks. Each core is sawn into two pieces along its longitudinal axis, and the sawn surfaces are then successively polished with No. 120 and 400 wet silicon carbide abrasive powders to make it suitable for microscopical observations (procedure similar to that described in ASTM C457). The observation area is usually limited to the top 20 mm (0.8 in.) of the specimen closest to the prepared surface because no cracking was found below this location.

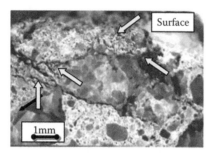

Figure 3.20 Optical microscope observation of surface preparation profiles obtained by jack-hammering (jack-hammer 14 kg). (From Bissonnette, B. et al., *Concr. Int.*, 28(12), 49, 2006.)

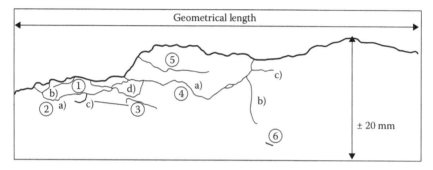

Figure 3.21 Typical cross section of a specimen observed under an optical microscope. (From Bissonnette, B. et al., *Concr. Int.*, 28(12), 49, 2006.)

Crack widths can be measured directly on the samples using a microscope and a typical field crack comparator with graduations at 0.05 mm (0.002 in.). For further characterization and comparison of the crack lengths, crack maps (Figure 3.20) can be digitalized (Figure 3.21) and observations performed by means of image analysis software.

3.5 POROSITY

Concrete inevitably contains pores with sizes ranging from the nanometer to millimeter scales. The pore structure deeply influences the properties of hardened concrete, notably those pertaining to its durability. Penetration of water, external species such as chloride and sulfate ions, and carbon dioxide occurs through the pore structure, and this is the main cause of concrete degradation (Wong and Buenfeld, 2006). Due to several phenomena happening during the hydration process and the excessive water-to-cement ratio (Neville, 2006), fluids are able to move into the porosity of the mortar phase of concrete by capillary absorption or permeation. As the NSL is usually more porous than bulk concrete, these fluid transfers are particularly important in this zone.

Transport phenomenon at the interface between the concrete substrate and a new material (repair or treatment) is conditioned by internal structure of solid material and characteristics of liquids or gases, such as viscosity, shear level, surface free energy, etc. Particularly, porosity, impregnation ratio, and water absorption have to be investigated to correctly define superficial concrete substrate. Open porosity is defined as the specific volume filled by water absorption and total porosity as the specific volume by water absorption under vacuum. Open and total porosity can be deduced from specific volumes. However, this test is a "bulk" test: water is coming into the sample by the six faces of the sample, independently of the quality of the surface in relation with the repair or surface material.

The significant parameters concern the distribution of the pores at the surface level. These properties can be evaluated by means of specific water absorption and water absorption under vacuum tests (Zajc et al., 2006), completed by mercury porosimetry intrusion (Courard et al., 2003). Calculation of porous volume as well as pore size distribution is based on these complementary experiments to estimate the quantity of water that is absorbable by the concrete.

The most commonly used test to analyze water transfer at the interface is the capillary suction test (Justnes, 1995; Courard et al., 2003). Mass change is usually measured after 5, 15, 30, and 45 min, as well as after 2, 6, and 24 h (EN 13057). Mass is measured on samples wiped off with a damp tissue. But, for investigating fluid transfer in the NSL, it is needed to record mass variation from the onset of water contact, which is why a specific testing procedure has been developed (Figure 3.22). Results presented in Figure 3.23

Figure 3.22 Continuous water absorption test setup. (From Courard, L. and Nélis, M., *Mag. Concr. Res.*, 55, 355, 2003a.)

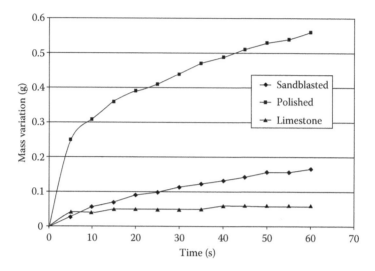

Figure 3.23 Water capillary suction between 0 and 60 s on (a) sandblasted, (b) polished concrete surface, and (c) limestone. (From Courard, L. and Degeimbre, R., *Canad. J. Civil Eng.*, 30(6), 1101, 2003b.)

show a major difference between specimens: mouth shape induced by surface preparation may explain the difference between sandblasted and polished concrete surface, while the very low porosity of limestone contributes to a reduced absorption level (Courard and Degeimbre, 2003b).

From the capillary suction test, it is possible to calculate the rate of water absorbed by the capillaries (E_c, %) with regard to the water absorbed under vacuum (E_v, %). The relative impregnation ratio S_t, at time t_i, is then evaluated as follows:

$$S_t = \frac{E_c}{E_v} \cdot 100$$

This value is representative of the absorption rate of the substrate at any time with respect to complete filling of the porous network (corresponding to accessible porosity of concrete samples) (Table 3.9).

Table 3.9 Comparative impregnation rates of concrete surfaces and limestone

Sample	Impregnation rates (%/$\sqrt{s}$)
Sandblasted concrete	0.1935
Polished concrete	0.5518
Limestone	1.1218

Source: Courard, L. and Degeimbre, R., *Canad. J. Civil Eng.*, 30(6), 1101, 2003b.

Table 3.10 Comparative mercury-intrusion porosimetry characteristics of concrete and limestone

Sample	Specific surface (m²/g)	Total porous volume (cm³/g)	Mean pore radius (nm)
Sandblasted concrete	1.81	0.0458	165
Polished concrete	1.87	0.0462	136
Limestone	0.54	0.0058	12

Source: Courard, L. and Degeimbre, R., *Canad. J. Civil Eng.*, 30(6), 1101, 2003b.

Other techniques, like mercury intrusion porosimetry (MIP) and nitrogen adsorption isotherms, can yield a more accurate view of the internal skeleton of concrete; they concern low-radius capillaries, that is, between 2 nm and 75 μm, which are unable to absorb large quantities of water in a short period of time. Tests performed on 10-mm cubic specimens reveal significant differences between concrete and limestone (Courard et al. 2003). As can be seen in Table 3.10, the pore volume measured for limestone aggregate is approximately 8 times less than that of concrete and the average pore size is 12 times smaller. Mass transfer will thus be influenced by the relative amount of paste and aggregates in the concrete skin.

Test procedures have been developed to measure the air and water permeabilities and the water absorption (sorptivity) directly on site. In this last case, the volume of water penetrating into the concrete is recorded at a constant pressure of 0.2 bars. As the water introduced by a syringe is absorbed by capillarity, the pressure inside—under the piston—tends to decrease; hence, it is maintained constant by pushing the piston through the cylinder (Figure 3.24). A plot of the quantity of water absorbed and the square root of time elapsed is linear, and the slope of this graph represents the sorptivity index (m³/min).

The same type of test device is based on a watertight gasket glued on the concrete substrate (Figure 3.25). The valves of the housing are opened, and the chamber is filled with boiled water. The valves are closed, and the top lid of the housing turned until a desired pressure is achieved (between 0 and 6 bars). The pressure selected is maintained with a micrometer gauge, pressing a piston into the chamber substituting the water penetrating the concrete.

The RILEM tube test method is also a standard water absorption test performed under low pressure in conformity with the method outlined in 11.4 (RILEM, 1980). It consists of a graduated glass tube, as shown in Figure 3.26, which is sealed against the surface under test with commercial modeling clay or putty. During the test, the tube is filled with distilled water to a known level. The water level is measured after 5, 10, 15, 20, and 30 min.

A specific system has been used (Courard et al., 2011) for evaluating the porosity of the "skin" concrete. With this aim, a specific test method has

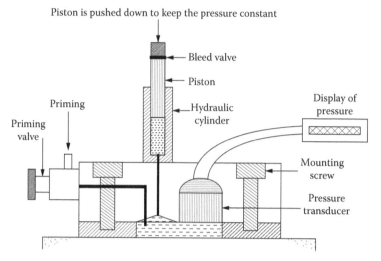

Figure 3.24 In situ permeability test—Method operation for water flow tests. (From Courard, L. et al., Condition evaluation of the existing structure prior to overlay, Chapter 3, 193-RLS RILEM TC Bonded cement-based material overlays for the repair, the lining or the strengthening of slabs and pavements, RILEM STAR Report Volume 3, 193-RLS RILEM TC, Springer, Dordrecht, the Netherlands, 2011, pp. 17–50.)

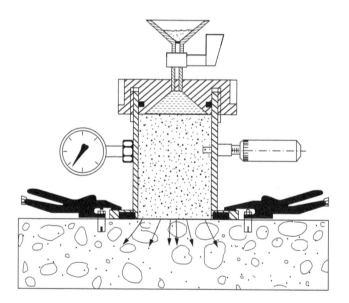

Figure 3.25 Water permeation test. (From Courard, L. et al., Condition evaluation of the existing structure prior to overlay, Chapter 3, 193-RLS RILEM TC Bonded cement-based material overlays for the repair, the lining or the strengthening of slabs and pavements, RILEM STAR Report Volume 3, 193-RLS RILEM TC, Springer, Dordrecht, the Netherlands, 2011, pp. 17–50.)

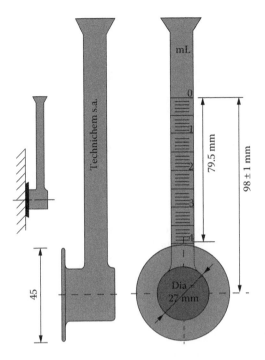

Figure 3.26 Kersten pipe system.

been set up: the "initial surface absorption test (ISAT)," developed on the base of Queen's University of Belfast testing device (Courard et al., 2011b). Autoclave (Figure 3.27) is a testing system for air and water permeability measurement of concrete (CNS Electronics). It can be used in the lab as well as on site; a metallic ring is fixed on the concrete substrate (Figure 3.28) and

Figure 3.27 Autoclave system and electronic controller. (From Courard, L. et al., *Constr. Build. Mater.*, 25(5), 2488, 2011.)

Figure 3.28 Gluing the ring onto the concrete substrate (ISAT). (From Courard, L. et al., *Constr. Build. Mater.*, 25(5), 2488, 2011.)

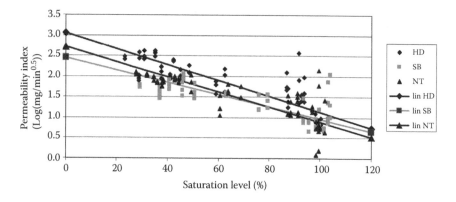

Figure 3.29 Permeability index vs. degree of saturation (ISAT). (From Courard, L. et al., *Constr. Build. Mater.*, 25(5), 2488, 2011.)

the quantity of the fluid—air or water—that is pushed into the concrete for a classical pressure of 0.5 kg/cm² is recorded continuously. The rate of the linear curve between 5 and 15 min classically determines a permeability index.

Good correlation between the permeability index and the degree of saturation has been observed (Figure 3.29); the higher the saturation, the higher the permeability index.

The degree of saturation has however a clear effect on capillary absorption (Figure 3.30). For surface preparation effect, the same conclusions can be given: hydro-jetting induces a higher rate of water capillary absorption, which probably can be correlated to soft microcracking. Whatever the test used, good correlations are observed (Figure 3.31) between the two methods.

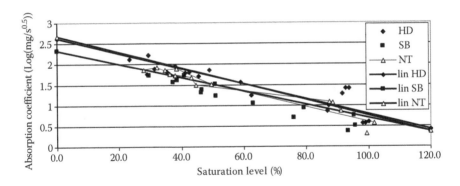

Figure 3.30 Capillary absorption rate versus degree of saturation (MCST). (From Courard, L. et al., *Constr. Build. Mater.*, 25(5), 2488, 2011.)

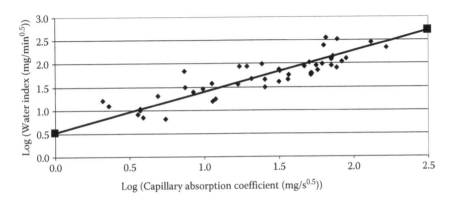

Figure 3.31 Comparison between permeability index and capillary absorption. (From Courard, L. et al., *Constr. Build. Mater.*, 25(5), 2488, 2011.)

3.6 MOISTURE CONTENT

3.6.1 In situ techniques

The most widespread test procedures available in North America to evaluate the moisture condition of concrete are described in ASTM E1907, "*Standard Guide to Methods of Evaluating Moisture Conditions of Concrete Floors to Receive Resilient Floor Coverings*" (ASTM E1907, 2006 [superseded]), as summarized in Table 3.11. Among the tests that are listed, some are standard test procedures, while others are not, and they are either classified as "qualitative" or "quantitative" procedures. The standard also provides recommendations for test frequency, location, and environment.

Electrical impedance testing relates the moisture content to the electrical AC impedance measured at the concrete surface. Several different

Table 3.11 Test procedures listed in ASTM E1907

Test method	Applicable standard	Results
Electrical impedance test	Nonstandard	Quantitative
Electrical resistance test	Nonstandard	Quantitative
Relative humidity test	ASTM F2170 (ASTM F2170, 2002)	Quantitative
	ASTM F2420 (ASTM F2420, 2005)	
Calcium chloride test	ASTM F1869 (ASTM F1869, 2004)	Quantitative
	Nonstandard	Qualitative
Plastic sheet test	ASTM D4263 (ASTM D4263, 1983 (2005))	Qualitative
Mat test	Nonstandard	Qualitative
Adhesive strip test	Nonstandard	Qualitative

Source: ASTM E1907-06a, Standard guide to methods of evaluating moisture conditions of concrete floors to receive resilient floor coverings, 2006.

moisture meters based on impedance measurement are commercially available. These nonstandard electronic test devices yield quantitative results, but in practice, it is generally considered as qualitative because it essentially provides indirect, comparative indications of moisture content in concrete (as relative weight %).

Electrical resistance test relates the moisture content (as relative weight %) to the measured electrical conductivity of concrete between the sensing probes or pins. The electrical resistance meters give a quantitative indication of the moisture content at the depth to which the probes or pins are inserted into the concrete surface. The main problem with these meters is that the electrical resistivity of concrete only partially depends on its moisture content. It depends on many other factors, such as degree of cement hydration, chemistry of hydration products, pH, carbonation, chlorides, etc.

In several countries, standards for concrete moisture measurements were developed based on R.H. measuring within a cavity inside the concrete element or at the surface. In the first approach, the probe is introduced in a cavity located at a given depth from the surface (cavity sealed on its walls, leaving only the bottom exposed within the enclosed volume) to measure the relative humidity inside the concrete element at that particular level. In the second approach, the probe is introduced in an enclosed space (hood) at the surface of the member to evaluate superficial relative humidity.

In both the ASTM F2170 (ASTM F2170, 2002) and the Scandinavian Nordtest method, in-place R.H. meters (Figure 3.32) are connected to a sensing probe that is inserted in a lined hole in the concrete. With most conventional relative humidity measuring devices, the main difficulty or hassle in performing R.H. measurement is related to the time required to take a single reading, as the probe generally needs to be left in the insulated cavity for quite a significant period of time before stabilizing. This is especially a problem when the number of locations to monitor on a given structure is important. To circumvent that problem, the probe can be left

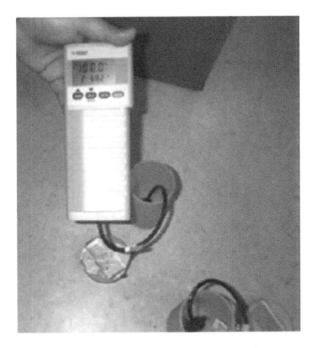

Figure 3.32 Relative humidity measurement in a concrete slab. (From ACI 302.2R, Guide for concrete slabs that receive moisture-sensitive flooring materials, American Concrete Institute, Farmington Hills, MI, 2006.)

in the cavity in-between recordings. The following types of R.H. meters, which can be mobilized throughout the monitoring period, are available at an affordable cost:

- Disposable electrical RH probes
- Wooden probes (electrical conductivity)
- Graphite probes

In the ASTM F2420 test procedure, the relative humidity prevailing at the surface of the concrete element is measured. Instead of a cavity located within the element, an insulated hood is sealed tightly against the surface at the test location, providing a small enclosed space that will equilibrate with the concrete surface relative humidity. This is the main difference with the ASTM F2170 procedure described previously. The relative humidity measured at the surface is in fact strongly linked with the *Vapor Emission Rate* measured with ASTM F1869 (ASTM F1869, 2004) procedure. In that respect, it must be emphasized that the moisture content at the surface can be significantly lower than that prevailing a few millimeters below.

The *Plastic sheet* test (ASTM D4263, 1983 (2005)) is relatively simple (Figure 3.33), but its reliability is somewhat questionable. As an example,

Figure 3.33 ASTM D4263 *Plastic Sheet* test—accumulation of water underneath the plastic causes the surface to feel cooler and often results in a darker color. (From ACI 302.2R, Guide for concrete slabs that receive moisture-sensitive flooring materials, American Concrete Institute, Farmington Hills, MI, 2006.)

Surprenant (2003) reported the results of two plastic sheet tests being conducted beside a calcium chloride test (ASTM F1869, 2004) for comparison testing. The plastic sheet tests showed no evidence of moisture, while calcium chloride tests conducted next to the sheets yielded emission values as high as 6.3 kg/100 m²/24 h.

The calcium chloride MVER (ASTM F1869, 2004) value is itself affected to a significant degree by ambient temperature and relative humidity. In fact, both the plastic sheet and calcium chloride test results can be affected by the quality of concrete surface preparation.

The ASTM F2170 (ASTM F2170, 2002) relative humidity test results are not affected by the test environment and concrete surface preparation because they are taken with probes inserted in the holes. However, the R.H. test results are affected by the water/cement (W/C) ratio and the alkalinity level of the concrete. Kanare (2005) showed the moisture contents of concrete mixtures with different W/C may be identical even though the measured internal R.H. values differ. Conversely, for a given recorded internal R.H. value, the moisture contents of different concrete mixtures can be dissimilar.

Commercially available moisture meters are useful for making a quick survey to determine where to place quantitative moisture tests. However, they should be used with caution because the calibration curves provided by the manufacturers are not necessarily produced at a depth that is of interest for a particular application.

Considering the limitations of existing test methods, more than one moisture test method should ideally be employed to determine more reliably the moisture condition of a concrete surface. In ASTM F710 (ASTM F710, 2008), two moisture criteria are suggested for concrete floors to be covered with resilient flooring when no manufacturer's indications are provided, based respectively on the MVER test (ASTM F1869, 2004) and the relative humidity test (ASTM F2170, 2002):

- ASTM F1869 (MVER test): MVER < 170 µg/m^2 (per 24 h period)
- ASTM F2170 (surface R.H. test): R.H. < 75%

Other than the practical test procedures summarized in Table 3.11, sophisticated *nuclear moisture* meters are also available for on site concrete moisture measurement. These portable devices (Figure 3.34) feature a radioactive source that emits gamma rays and high-speed neutrons. The neutrons are slowed by interactions with hydrogen atoms in concrete and water, being converted into "thermal" neutrons, which are backscattered and detected by a sealed gas counter in the instrument. A digital display on the instrument indicates the number of counts collected over a fixed time period, generally 10–60 s per measurement series. This type of instrument can provide useful information on relative differences in moisture conditions to a depth of 100 mm.

To ensure adequate use of moisture testing data, a thorough understanding of the test methods and their limitations is necessary. Results from moisture testing should be interpreted, taking into account that

- Moisture is not static, it migrates within the concrete porosity according to the boundary conditions.
- During concrete drying, the surface moisture content is normally lower than the moisture content in the bulk concrete.
- Relative humidity measurements taken with a surface-mounted hygrometer are normally lower than measurements taken with a R.H. probe embedded in concrete, even at shallow depths.

Existing standard procedures for concrete moisture evaluation were essentially developed for the concrete floor slabs covering industry. There are still many limitations to measuring the moisture of concrete, and at this stage, it is not yet possible to draw a definite conclusion on universally reliable testing method(s) for rapid moisture content assessment.

Figure 3.34 Nuclear moisture meter for the measurement of relative humidity in concrete. (From ACI 302.2R, Guide for concrete slabs that receive moisture-sensitive flooring materials, American Concrete Institute, Farmington Hills, MI, 2006.)

3.6.2 Laboratory techniques

This section addresses some test methods specifically designed to measure the moisture content or relative humidity of concrete samples in the laboratory.

Using a *thermogravimetric moisture content evaluation* approach, the free water content in concrete can be determined for representative samples extracted from a concrete laboratory specimen or in situ element. Ideally, the test sample should be a cored cylinder with a diameter equal to at least three times the aggregate top size. The test specimen should be dry-cut to avoid modifying the actual moisture content with water from the core extraction operation. Alternatively, specimens of concrete can be stitch-drilled and chiseled from the concrete structure or element. The concrete sample is weighed and then heated at 105°C until the weight loss is not more than 0.1% in 24 h of oven-drying. The weight loss is calculated and expressed as a percentage of the dry weight. This technique can yield a very accurate measurement of the actual moisture content in concrete, provided that the specimen preparation is performed with adequate care.

Nuclear magnetic resonance (NMR), which is based on the interaction of a transmitted radio frequency field and nuclear particles (hydrogen nuclei) contained in a static magnetic field, can be used to determine the amount and binding state of a hydrogen-bearing liquid, water for instance, in a porous medium like concrete. Although conventional NMR is *per se* a non-destructive method, sampling is required in the case of a large specimen. *One-Side Access Nuclear Magnetic Resonance* (OSANMR) is used to measure the one-dimensional water distribution (moisture profile) in building materials (Kanare, 2005). An OSANMR device can be applied to the specimen surface from one side, which allows information about water transport during capillary absorption and pressure-driven permeation to be determined in situ.

3.7 SURFACE COMPOSITION

Owing to various factors discussed in the previous sections, the concrete skin composition may differ considerably from the bulk material composition. Its characterization is thus important, both qualitatively and quantitatively, notably with respect to aesthetic considerations and adhesion of repairs and surface treatments.

A specific procedure where each constituent at the interface is inventoried (Figure 3.35) has been developed by Courard (2005). Specimens are first sawn, polished, and impregnated with fluorescent resin. The impregnated face of each specimen is then abraded and polished until the aggregates appear. The counting of aggregates (elements ranging from 1.2 to 5 mm in size) is performed with a binocular microscope, while the evaluation of the cement paste proportion in the remaining portion is based on polarized microscope observations. Counting of elements greater than 1 mm is performed under the microscope, using a ruler graduated from 1.2 to 5 mm.

Counting is performed exclusively on polished concrete surfaces in order to avoid problems due to variable plane depths. Figure 3.36 shows an example of aggregate size distribution found at the concrete surface in the 1.2–5 mm size range (Courard, 2005). Based on the collected data, it could be concluded that 12% of the total surface was occupied by aggregates.

The counting of elements smaller in size than 1.2 mm (actually 1.125 mm) is performed on photographs observed through a polarized microscope. The areas where aggregates are found are blackened, and counting is carried out by image analysis, based upon color discrimination and pixel counting.

In the investigation reported by Courard (2005), the relative surface area really occupied by aggregates under 1.2 mm was estimated to amount to approximately 60% of the remaining portion of the surface (after removal of the elements equal to or greater than 1.2 mm).

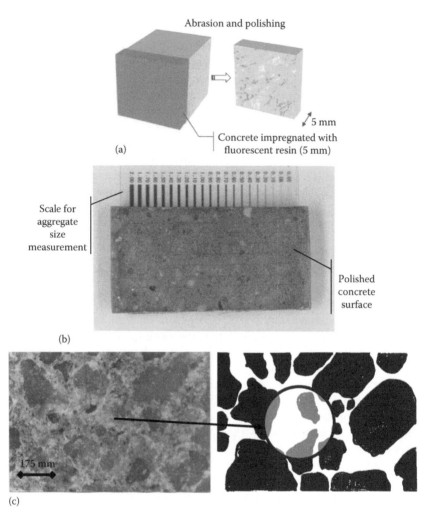

Abrasion and polishing

5 mm

Concrete impregnated with
fluorescent resin (5 mm)

(a)

Scale for
aggregate
size
measurement

Polished
concrete
surface

(b)

175 mm

(c)

Figure 3.35 Procedure for the observation of the topography of the concrete surface: (a) preparation of the samples, (b) counting of the aggregates greater
than 1 mm, and (c) counting of the aggregates inferior to 1 mm. (From
Courard, L., *Mag. Concr. Res.*, 57(5), 273, 2005.)

The relative surface area occupied by the aggregates is obtained by add-
ing the two aggregate fractions $12.08 + 87.98 \times 0.6 = 64.88\%$, in the example provided. This means that the cement paste and the aggregates covered
respectively 35% and 65% of the surface.

In reality, it has been seen before (§ 3.2) that cement paste borders
around aggregates are areas of the paste that are significantly more porous,
thereby likely to enhance mechanical anchorage as a result of more extensive cement paste (or resin) intrusion. In evaluating the importance of this

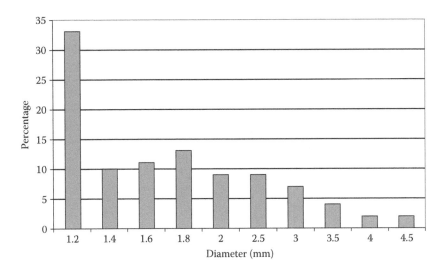

Figure 3.36 Distribution of the aggregate sizes at the surface of concrete (1.2/4.5 mm). (From Courard, L., *Mag. Concr. Res.*, 57(5), 273, 2005.)

Interfacial Transition Zone, it was found that it actually represents an important fraction of the cement paste. The following proportions are typical of what is found on a concrete surface where the finished or formed surface layer has been removed (Figure 3.37):

- 65% aggregates and sand particles
- 2% cement paste
- 33% ITZ

3.8 CHEMICAL CONTAMINATION

Deleterious chemical species may be present in the NSL due to environmental or internal contamination. Depending on many factors, it may induce the loss of properties for the concrete surface, often as a result of expansion and cracking (see Section 3.4) of the cement-based structure (Courard et al., 2010). The most common contaminants affecting concrete structures include:

- Sulfate and nitrate attacks: soils, polluted water, sea water, chemical pollutants, etc.
- Chlorides: deicing salts, chemical pollutants, etc.
- Carbonation
- Hydrocarbons

The assessment of concrete must be based on the cartography of the degradations in the plane and the depth of the concrete: what are the zones

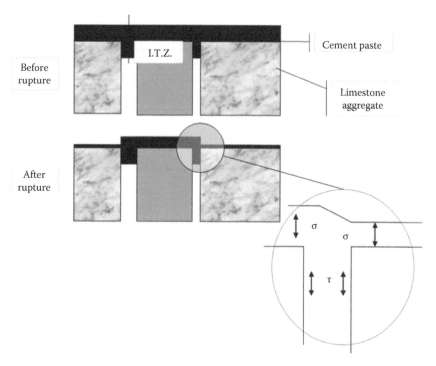

Figure 3.37 Influence of ITZ as observed on the interface between cement paste and concrete substrate, before and after bond strength test. (From Courard, L., *Mag. Concr. Res.*, 57(5), 273, 2005.)

affected and what is the profile of the species concentrations on the depth of the concrete structure. This usually needs the coring of samples and the evaluation of the affected concrete by means of classical analytical chemistry as well as microscope observations (polished sections, thin sections, and SEM). These investigations can lead to

- *Polished sections*: Type of aggregates and sand, distribution of aggregates and sand, cracking, depth of carbonation, type of reinforcement and concrete cover, layer sequence and thickness, etc.
- *Thin sections (Figure 3.38)*: W/C ratio, type of cement/binder, risk of AAR, frost resistance, carbonation, influence of moisture transport, defects from casting/curing, etc.
- *Scanning electron microscope with EDX*: Atoms identification, porosity, hydrated cementitious species, special mineralogy elements, etc.

Sulfate and nitrate attacks are specially expected in special environments like sewage plants, agricultural floors, marine and food infrastructures, etc. Depending on the concentration (Darimont et al., 2006), they lead to

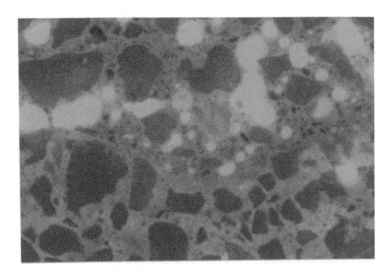

Figure 3.38 Interface between concrete substrate and repair material (with a lot of air bubbles). (From Courard, L. et al., Repairing concrete with self compacting concrete: Testing methodology assessment, in: Shah, S.P., ed., *First North American Conference on the Design and Use of Self-Consolidating Concrete (SCC)*, November 12–13, 2002, Rosemont, Chicago, IL, Center for Advanced Cement-Based Materials, 2002, pp. 267–274.)

(a)   (b)

Figure 3.39 (a) Secondary ettringite and (b) calcium chloroaluminate. (From Darimont, A. et al., *Bull. des laboratoires des Ponts et Chaussées*, 261–262, 121, 2006.)

the progressive disintegration of the cement paste due to expansion reactions like secondary ettringite and calcium chloroaluminates (Figure 3.39).

The depth of carbonation is simply assessed with the phenolphtalein test, either on site or in the laboratory. This will establish the relationship between depth of carbonation and cover in various areas of the structure for a given age, in order to define areas of risks.

Chloride ion content can be determined rapidly and cheaply from the analysis of drilled dust samples taken from the concrete. However, chloride contents can be highly variable around a structure. Measurement of half-cell potential can be used to find suspect areas, from which dust drilling samples can be taken.

There is a common perception that hydrocarbons affect the setting and hardening of the concrete, giving a reduced long-term strength. When hardened concrete is into contact with oil, kerosene, and other hydrocarbon materials, penetration is possible in the NSL (Wilson et al., 2001). The loss of mechanical properties (compressive and splitting tensile strengths), resulting from exposure to oil, is relatively smaller for high performance concrete compared with normal strength concrete (Jasim and Jawad, 2010). But the effects of petroleum hydrocarbons on hardened concrete, which has achieved its design strength, are of limited concern. Creosote, however, can affect hardened concrete that has achieved its design strength. Where hardened concrete is likely to come into contact with creosote-derived contamination, then a reduction in long-term strength should be considered (Wilson et al., 2001).

Determination of hydrocarbon content into concrete may be performed by means of infrared spectrometry on leachates extracted from the NSL.

3.9 AESTHETICS

The external aspect of concrete surfaces influences the appreciation we have for the appearance of the entirety of the structure and will indirectly imply our judgment about the wellness induced by material (Lemaire et al., 2005). The external aspect of concrete surface is usually estimated based on subjective and objective parameters. Aesthetic requirements of architects and building owners with regards to the quality of concrete surfaces become however more and more stringent. But contractors and concrete makers are often incompetent to select materials and propose solutions to fulfill these requirements. They need methods, guides, and objective standards for helping them designing these specific concretes (Pleau et al., 1999).

The development of tools to characterize the color of concrete surfaces and the study of the effects of various parameters influencing the color (Figure 3.40) are just starting to emerge (Lallemant et al., 2000; Lemaire et al., 2005). Some researchers have proposed a scientific approach to assess color as a measurable aesthetic property of concrete (Thompson, 1969). Yet, the aesthetic qualities of concrete surfaces are also related to their texture (in terms of roughness and lightness), which depends on the chemical and physical properties of the concrete skin at the microscopic scale (Lallemant et al., 2000; Martin, 2007). Moreover, other parameters also need to be mentioned, including type and quality of materials, water/cement (W/C) ratio, environmental conditions of working, casting technology, curing and, finally, formwork material (Martin et al., 2008).

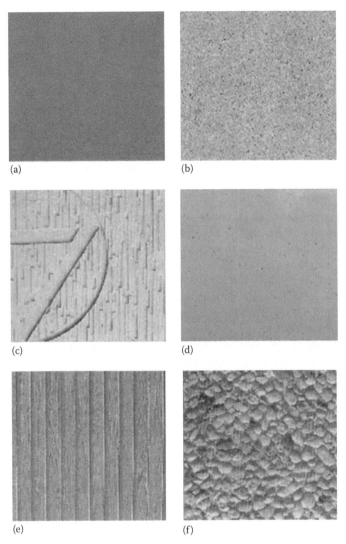

Figure 3.40 Samples of the possible rendering of concrete surfaces: (a) mold finishing: smooth and dark, (b) acid-etched, (c) mold finishing: textured, (d) mold finishing: smooth and light, (e) mold finishing: textured, and (f) water-washed. (Adapted from FeBe, *Memento of Architectonic Concrete*, 2nd edn., Belgian Federation of Concrete Industry, Bruxelles, Belgium, 1996, in French.)

Concrete surface aesthetism can be characterized by color and texture. The appraisal of the aesthetic quality of a concrete surface is generally performed by rating the various areas based on a color scale and a blowhole scale (Figure 3.41). For color evaluation, a colorimeter can alternatively be used. The results depend on the operator; moreover, it is difficult to appreciate the color of an entire surface with local measurements.

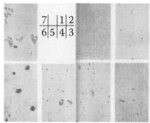

Figure 3.41 Gray scale and chart for blowholes. (From CIB Report No. 24, Tolerances on blemishes of concrete, Document established by commission W29 of CIB, 1973.)

When architects design concrete structures and elements, they have to select the type of finish and prepare adequate specifications, notably with regard to quality control, to ensure that their objectives are met satisfactorily. Actual standard and technical documents in Europe define concrete surface classification based on the extent of defects oberved over the concrete surface. All these documents are based on the CIB report (CIB Report, 1973). The CIB report classifies concrete surfaces into four categories and defines for each category a maximum level of defect. The defects considered in the document are flatness, stain and local surface defects, holes (blowholes), and tint variation, respectively:

• Flatness is defined by the maximum deflection on the concrete surface; a maximum flatness is defined for each class of concrete facing.
• Stain and local defects characterized by the maximum area of defect surface defined use function of distance between concrete and observer. For each class of concrete, surface tolerances are given.
• Holes are characterized by an appreciation of blowholes concentration distributed over the entire surface. Concentration of blowholes is made by comparison between the concrete facing and a seven-reference scale. For each concrete facing class, a maximum level of holes is defined.
• Maximum tint variation is evaluated by comparing between concrete facing and a gray scale. For each concrete facing class, an acceptable deviation in tint variation is defined.

The French and Belgian Standards are close to the CIB report (CIB Report, 1973). These standards do not contain provisions for detailed specification of concrete surface characteristics. The Luxembourgian Standard CDC-BET gives more information for concrete surface specification, which includes precautions on materials selection and storage (CDC, 2007). It also defines surface defects that can be avoided. This document clearly explains that concrete surface quality depends on the formwork characteristics and the materials used.

Special attention must be paid to these considerations by contractors whenever a special concrete surface has to be produced.

Based on the documents, requirements for concrete include surface classification, concrete tint, and accepable amount of holes. Surface concrete tint and allowable holes are addressed in the CIB document. However, the provisions are not adapted for colored concrete, and in practice, it can be difficult to objectively compare an entire surface with a pocket-size gray scale comparator. Similarly, it is difficult to evaluate blowholes over an entire surface by comparing with a small-size comparator.

Upon delivery inspection, to check the conformity with the blowhole requirement, the seven-level scale is placed onto the concrete surface, and the inspector must be standing at a distance between 3 and 10 meters, depending on the standard used. The observer will compare the seven-reference scale with the concrete facing and identify the blowhole level of the concrete surface. This method is subjective and does not yield important information like the percentage of surface exhibiting holes, the estimated number of holes, and the hole size range.

To check the conformity with the color requirements, the first step is to appraise the surface tint variations with the gray scale, like in the blowhole verification procedure. Alernatively, to appraise the surface tint on site, photographs may be used. In this case, the gray scale is fixed onto the concrete surface. The surface should be photographed in black and white, at the required distance. Photographs should be developed on matt paper and in sizes that enable distinguishing the levels of the gray scale. The assessment consists of two analyses carried out on the photo by three persons. The color of the surface should first be identified with regard to the gray scale of the photograph. Then a second assessment should be carried out, placing the gray scale directly on the photo. In case of disagreement, the assessment should be carried out as follows: color measurement should be performed with a colorimeter in L, a, and b coordinates. Such analysis should be carried out evenly over the disputed part of the surface. The actual color mapping should be performed and compared to the allowable tint variations.

These procedures can be applied when repairing or coating concrete surface. As already mentioned, it is difficult to guaranty the aesthetic quality of the work in advance, which is why mockups are usually required if aesthetic requirements are to be involved with physical or mechanical requirements.

3.10 OTHER PROPERTIES

In the previous sections, test procedures for the evaluation of the most important properties of a concrete surface with respect to repair and surface treatment were described. In some applications, other surface properties have to be assessed. For instance, as stated in EN 1504-10, an evaluation of the surface evenness is recommended prior to applying protective coating or strengthening with reinforcing plates; evaluation of surface resistivity is required in the case of repairs where electrochemical techniques are used, notably realkalization of carbonated concrete, chloride ion extraction, and cathodic protection. It should be stressed however that these properties refer to the state of surface after repair or treatment of the concrete structure.

Examples of this kind of approach can be found for screeds (EN 13813) applied on concrete substrate. In the case of cement screeds, the following surface properties need to evaluated:

- *Normative*: Wear resistance (using one of the three recommended tests)
- *Optional*: Surface hardness, wear resistance to rolling wheel, impact resistance

3.10.1 Surface evenness

More often, the evenness evaluation is used in the case of repaired or protected surfaces, for example, industrial floors. According to EN 1504-10, surface evenness of substrate prepared for repair as well as after application of mortars should be visually evaluated. For this purpose, laser gauges and straightedges may be also used (Zajc et al., 2006). A private company developed a profilometer which can be also used for fast evaluation of concrete surface evenness (Figure 3.42).

3.10.2 Electrical resistivity

The electric resistivity of concrete is used for indirect evaluation of concrete characteristics such as permeability and properties of the pore water solution, namely for characterizing chloride ion diffusivity. Theoretical and experimental studies indicate a correlation between concrete resistivity and chloride ingress (Gulikers, 2005). ASTM C1202 standard test method and AASHTO T277 are usually recommended for the assessment

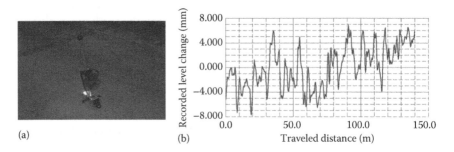

(a)

(b)

Figure 3.42 Assessment of surface evenness in concrete floors: dipstick device (a); and example of measurements (b). (From Warsaw University of Technology, unpublished results.)

Figure 3.43 Example of a device for evaluation of concrete electrical resistivity with the Wenner test (www.proceq.com).

of concrete resistivity to chloride penetration. However, a simple non-destructive surface resistivity test has been developed. The *Wenner test* was implemented for testing concrete properties (Figure 3.43). Originally, the Wenner test was used for measurement of soil resistivity and the estimation of expected corrosion rates and for the design of cathodic protection systems (ASTM G57-06).

According to EN 1504-10, it shall be used for evaluation of electric resistivity of concrete substrate prepared prior to repair as well as electrical properties of an applied repair mortar.

The resistivity ρ of the concrete slab measured (thickness $\gg a$) with a Wenner probe can be calculated from the formula:

$$\rho \approx \rho_{\text{app}} = 2\pi a \frac{P'}{I'} \tag{3.5}$$

where

ρ is the resistivity of concrete

ρ_{app} the apparent value of resistivity (the value shown in the display screen of commercial probes)

a is the distance between consecutive points of the array

P' the potential drop between the electrodes at a current I' when applied to a finite body

Most of the tests are performed on standard cylindrical samples taken from concrete structures (Morris et al., 1996; Ramezanianpour et al., 2011).

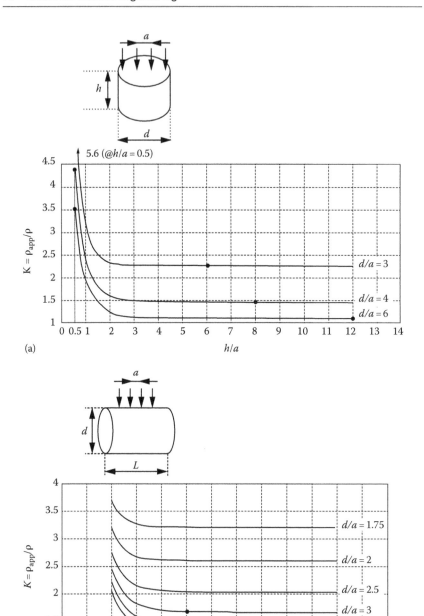

Figure 3.44 Cell constant correction to determine the concrete resistivity for centered: (a) end face and (b) longitudinal configurations. (Adapted from Morris, J. et al., *Cement Concr. Res.*, 26(12), 1779, 1996.)

The effects of a size sample and a test configuration should be taken into account for results interpretation. The formula for concrete resistivity should be modified by cell constant correction coefficient K, which is the function of interprobe distance and sample geometry (Figure 3.44b):

$$\rho = \frac{\rho_{app}}{K} \tag{3.6}$$

The experimental investigations performed by Ramezanianpour et al. (2011) on concretes of various compositions indicate a very good correlation between surface resistivity SR and RCTP results of chloride (Figure 3.45a) and water (Figure 3.45b) penetration. The relation between SR and compressive strength of concretes was less statistically significant. This can be explained by higher sensitivity of electrical resistivity to microstructure changes than compressive strength (Figure 3.45c). Morris et al. also showed the effect of a grain size on the value of resistivity measure with Wenner test (Figure 3.44).

3.10.3 Wear resistance

Wear resistance is important for the service life of various concrete structures such as pavements, concrete highways, airfield runways, parking lots, hydraulic structures, tunnels, dam spillways, etc. Deterioration of surfaces of these structures occurs due to various forms of wear such as erosion (wearing by abrasive action of fluids containing suspended solids), cavitation (wearing by implosion of vapor bubbles in high-velocity fluid flow), and abrasion (wearing by repeated rubbing or frictional processes) (Horszczaruk, 2008; Genzel et al., 2011). The wear itself is difficult to define, and it originates from multiple sets of complex interactions on a microscopic scale between surfaces that are in mechanical contact and slide against each other (Figure 3.46). A large number of studies indicated that concrete abrasion resistance is primarily dependent upon the compressive strength of the concrete. Factors such as cement content, W/C ratio, type of aggregate, and their properties affect concrete strength. It was shown that wear resistance depends on the amount and hardness of aggregate used (Laplante et al., 1991). It is profitable to use crushed aggregates with a high hardness and a large amount of particles with a larger diameter located predominantly at the surface of the concrete element (Neville, 2012).

The wear resistance (Figure 3.47) for cementitious materials and synthetic resin materials, to be used as wearing surfaces, shall be determined in accordance with one of the test methods described as follows:

- *EN 13892-3*: Wear resistance Böhme (Figure 3.47a) is designated by "A" (for *Abrasion*) followed by a number which means the abrasion quantity in $cm^3/50\ cm^2$, defining seven classes of wear resistance (A22, A15, A12, A9, A6, A3, A1.5).

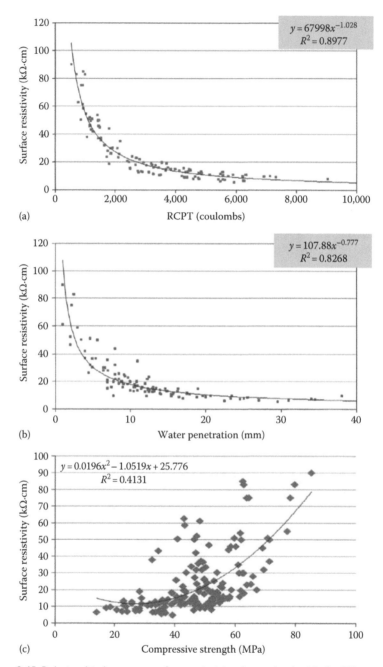

Figure 3.45 Relationship between surface resistivity determined with the Wenner test
        and: (a) RCPT results, (b) water penetration, and (c) compressive strength
        for concrete of various composition. (From Ramezanianpour, A.A. et al.,
        Constr. Build. Mater., 25, 2472, 2011.)

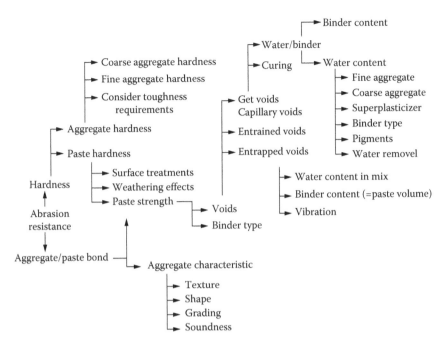

Figure 3.46 Various parameters affecting abrasion resistance of concrete. (From Gencel, O. et al., *Indian J. Eng. Mater. Sci.*, 18, 40, 2011.)

- *EN 13892-4*: Wear resistance BCA (Figure 3.47b) is designated by "AR" (for *Abrasion Resistance*) followed by a number which means the maximum wear depth in 100 μm, defining five classes of wear resistance (AR6, AR4, AR2, AR1, AR0.5).
- *EN 13892-5*: Wear resistance to rolling wheel (Figure 3.47c) is designated by "RWA" (for *Rolling Wheel Abrasion resistance*) followed by a number which means the abrasion quantity in cm^3, defining five classes of wear resistance (RWA300, RWA100, RWA20, RWA10, RWA1).
- *EN 14157* (Figure 3.47d): A test method developed for testing of stone materials can be also used for evaluation of abrasion of concrete precast elements, the so-called Capon test or "broad disk" method.

These test methods are designed to evaluate the dry-friction abrasion resistance of concrete. A few publications and ASTM C 1138-97 describe research done in conditions similar to the natural influence of environment with the applications of devices enabling to model the process of concrete abrasion with the mixture of aggregate and water. Horszczaruk (2004) has designed a special device for testing erosion (Figure 3.48). The wear test with this device confirmed the influence of concrete quality also on wear resistance in abrasive erosion conditions (Figure 3.49).

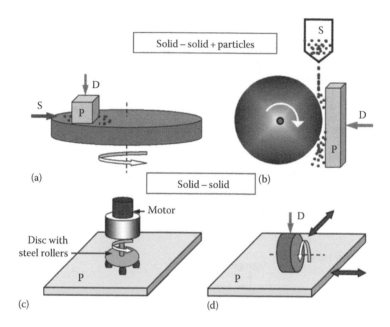

Figure 3.47 Scheme of the most common test methods for evaluation of concrete wear resistance: (a) Böhme test, (b) "broad disk" (Capon test), (c) BCA wear test, and (d) steel wheel for testing rolling wheel abrasion; D—normal load direction, P—concrete sample, S—abrasive. (From Horszczaruk, E., *The Model of Abrasive Wear of Cement Concrete*, Publishing House of the West-Pomeranian University of Technology, Szczecin, Poland, 2008.)

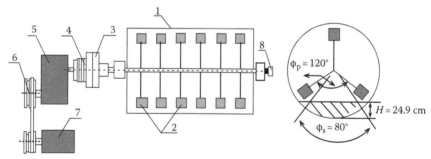

1—drum, 2—concrete samples, 3—flexible coupling, 4—torque meter with collector, 5—reduction gear, 6—belt transmission, 7—electric motor, 8—rev-counter, H—abrasive mix level, ϕ_p—angle between arms, ϕ_s—angle of contact with abrasive mix

Figure 3.48 Scheme of the device for testing abrasive erosion of concrete. (From Horszczaruk, E., *Wear*, 256, 787, 2004.)

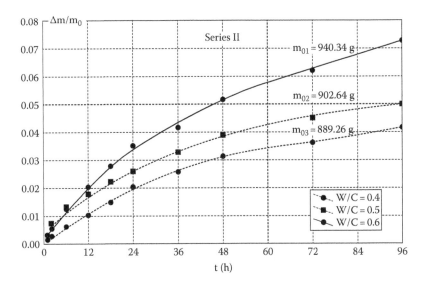

Figure 3.49 Abrasion resistance test results for concretes with different W/C ratio. (From Horszczaruk, E., *Wear*, 256, 787, 2004.)

3.11 CONCLUSION

Proper characterization of the concrete substrate, especially the near-to-surface layer, is essential for reliably achieving strong adhesion of repair materials and surface treatments. The chemical, physical, and mechanical make up of the surface at the time of the intervention influence the thermodynamic conditions governing the development of interfacial bond. In the framework of concrete surface engineering, the surface characterization techniques presented in this chapter allow to determine the various properties and characteristics that are of relevance in a comprehensive analysis of adhesion mechanisms and bond development.

REFERENCES

AASHTO T277-07 (2011) Standard method of test for rapid determination of the chloride permeability of concrete. Washington, DC: American Association of State Highway and Transportation Officials.

ACI 302.2R-06 (2006) Guide for concrete slabs that receive moisture-sensitive flooring materials. Farmington Hills, MI: American Concrete Institute, 42pp.

ACI 440.2R-08 (2008) Guide for the design and construction of externally bonded FRP systems for strengthening concrete structures. Farmington Hills, MI: American Concrete Institute, 76pp.

ACI 503.1-92 (2003) Standard specification for bonding hardened concrete, steel, wood, brick and other materials to hardened concrete with a multi-component epoxy adhesive. Farmington Hills, MI: American Concrete Institute, 5pp.

ACI 562-13 (2013) Code requirements for evaluation, repair, and rehabilitation of concrete buildings and commentary, Farmington Hills (MI), American Concrete Institute, 59pp.

ANSI/ASME B46.1-1978 (2009) Surface texture, surface roughness, waviness and lay. New York: American Society of Mechanical Engineers.

ASTM C109 / C109M-13 (2013) Standard test method for compressive strength of hydraulic cement mortars (using 2-in. or [50-mm] cube specimens). West Conshohocken, PA: ASTM International.

ASTM C418-12 (2012) Standard test method for abrasion resistance of concrete by sandblasting. West Conshohocken, PA: ASTM International, 4pp.

ASTM C597-09 (2009) Standard test method for pulse velocity through concrete. West Conshohocken, PA: ASTM International, 4pp.

ASTM C805/C805M-13a (2013) Standard test method for rebound number of hardened concrete. West Conshohocken, PA: ASTM International, 4pp.

ASTM C873/C873M-10a (2010) Standard test method for compressive strength of concrete cylinders cast in place in cylindrical molds. West Conshohocken, PA: ASTM International, 4pp.

ASTM C944-12 (2012) Standard test method for abrasion resistance of concrete or mortar surfaces by the rotating-cutter method. West Conshohocken, PA: ASTM International, 5pp.

ASTM C1138-97 (1997) Standard test method for abrasion resistance of concrete (underwater method). West Conshohocken, PA: ASTM International, 4pp.

ASTM C1450 / C1450M-04 (2014) Standard specification for non-asbestos fibercement storm drain pipe, ASTM International. West Conshohocken, PA: ASTM International, 5pp.

ASTM C1202-12 (2012) Standard test method for electrical indication of concrete's ability to resist chloride ion penetration. West Conshohocken, PA: ASTM International, 7pp.

ASTM D4263-83 (2012) Standard test method for indicating moisture in concrete by the plastic sheet method. West Conshohocken, PA: ASTM International, 2pp.

ASTM E965-15 (2015) Standard test method for measuring pavement macrotexture depth using a volumetric technique. West Conshohocken, PA: ASTM International, 4pp.

ASTM E1907-06a (2006) Standard guide to methods of evaluating moisture conditions of concrete floors to receive resilient floor coverings (withdrawn 2008). West Conshohocken, PA: ASTM International, 9pp.

ASTM E2157-15 (2015) Standard test method for measuring pavement macrotexture properties using the circular track meter, ASTM International. West Conshohocken, PA: ASTM International, 5pp.

ASTM F710-11 (2011) Standard practice for preparing concrete floors to receive resilient flooring. West Conshohocken, PA: ASTM International, 8pp.

ASTM F1869-11 (2011) Standard test method for measuring moisture vapor emission rate of concrete subfloor using anhydrous calcium chloride. West Conshohocken, PA: ASTM International, 4pp.

ASTM F2170-11 (2011) Standard test method for determining relative humidity in concrete floor slabs using in situ probes. West Conshohocken, PA: ASTM International, 5pp.

ASTM F2420-05 (2005) Standard test method for determining relative humidity on the surface of concrete floor slabs using relative humidity probe measurement and insulated hood (withdrawn 2014). West Conshohocken, PA: ASTM International, 6pp.

ASTM G57-06 (2012) Standard test method for field measurement of soil resistivity using the Wenner four-electrode method. West Conshohocken, PA: ASTM International, 6pp.

Austin, S.A. and Robins, P.J. (1993) Development of a patch test to study the behaviour of shallow concrete patch repairs. *Concrete Research*, 45(164), 221–229.

Barker, M.G. and Ramirez, J.A. (1988) Determination of concrete strengths with break-off tester. *ACI Materials Journal*, 85(4), 221–228.

Bissonnette, B., Courard, L., Vaysburd, A., and Bélair, N. (2006) Concrete removal techniques: Influence on residual cracking and bond strength. *Concrete International*, 28(12), 49–55.

BS 1881-122 (2011) Testing concrete. Method for determination of water absorption. London, U.K.: British Standards Institution.

Bungey, J.H. (1989) *The Testing of Concrete in Structures*, 2nd edn. New York, NY: Chapman & Hall, pp. 94–110.

Bungey, J. H. and Madandoust, R. (1992) Factors influencing pull-off tests on concrete. *Magazine of Concrete Research*, 44(158), 21–30.

Bhushan B. (2001) Surface Roughness Analysis and Measurement Techniques, in: Modern Tribology Handbook (ed. B.Bhushan), CRC Press LLC, N.W. Corporate Blvd, Boca Raton, FL, 1760pp.

Carino, N.J. (2003) Non destructive test methods to evaluate concrete structures. *Sixth CANMET/ACI International Conference on the Durability of Concrete*, Special Seminar, Thessaloniki, Greece, 75pp.

CDC-BET (2007) Requirements for concrete casting. Ministry of Public Works, Grand-Duché de Luxembourg (in French).

CIB Report No. 24 (1973) Tolerances on blemishes of concrete, Document established by commission W29 of CIB, Rotterdam, the Netherlands.

Cleland, D.J. and Long, A.E. (1997) The pull-off test for concrete patch repairs. *Proceedings of the Institution of Civil Engineers: Structures and Buildings*, 122(11), 451–460.

Cleland, D.J., Naderi, M., and Long, A.E. (1986) Bond strength of patch repair mortars for concrete. *Proceedings of the International Symposium on Adhesion between Polymers and Concrete*, Aix-en-Provence, France (Ed. H.R. Sasse), Chapman & Hall, London, U.K., pp. 235–244.

Courard, L. (2005) Adhesion of repair systems to concrete: Influence of interfacial topography and transport phenomena. *Magazine of Concrete Research*, 57(5), 273–282.

Courard, L., Bissonnette, B., Garbacz, A., Vaysburd, A., von Fay, K., Moczulski, G., and Morency M. (2014) Effect of misalignment on pull-off test results: Numerical and experimental assessments. *ACI Materials Journal*, **111**(2), 153–162.

Courard, L., Bissonnette, B., and Trevino, M. (2011a) Condition evaluation of the existing structure prior to overlay (Chapter 3). 193-RLS RILEM TC Bonded cement-based material overlays for the repair, the lining or the strengthening of slabs and pavements. RILEM STAR Report Volume 3, 193-RLS RILEM TC, Springer, Dordrecht, the Netherlands, pp. 17–50.

Courard, L., Bissonnette, B., Vaysburd, A., Bélair, N., and Lebeau, F. (2012) Comparison of destructive methods to appraise the mechanical integrity of a concrete surface. *Concrete Repair Bulletin*, **25**(4), 22–30.

Courard, L. and Bissonnette, B. (2004) Essai dérivé de l'essai d'adhérence pour la caractérisation de la cohésion superficielle des supports en béton dans les travaux de réparation: Analyse des paramètres d'essai. *Materials and Structures*, **37**(269), 342–350.

Courard, L., Darimont, A., Degeimbre, R., Willem, X., Geers, C., and Wiertz, J. (2002) Repairing concrete with self compacting concrete: Testing methodology assessment. In: *First North American Conference on the Design and Use of Self-Consolidating Concrete (SCC)*, November 12–13, 2002, (Ed. S.P. Shah), Center for Advanced Cement-Based Materials, Rosemont, Chicago, IL, pp. 267–274.

Courard, L. and Degeimbre, R. (2003b) A capillary suction test for a better knowledge of adhesion process in repair technology. *Canadian Journal of Civil Engineering*, **30**(6), 1101–1110.

Courard, L., Garbacz, A., Schwall, D., and Piotrowski, T. (2006a) Effect of concrete substrate texture on the adhesion properties of PCC repair mortar. In: *International Symposium on Polymers in Concrete (ISPIC)*, April 2–4, 2006, University of Minho, Guimaraès, Portugal (Eds. J. Barroso de Aguiar, S. Jalali, A. Camoes, and R.M. Ferreira), pp. 99–110.

Courard, L., Garbacz, A., and Wolff, L. (2006b) Evaluation and quality assessment of industrial floors (Chapter 4). RILEM TC 184—IFE Industrial floors for withstanding environmental attacks, including repair and maintenance. RILEM Report 33 (Ed. P. Seidler), RILEM Publications S.A.R.L., Bagneux, France, pp. 59–89.

Courard, L., Lenaers, J.F., Michel, F., and Garbacz, A. (2011b) Saturation level of the superficial zone of concrete and adhesion of repair systems. *Construction and Building Materials*, **25**(5), 2488–2494.

Courard, L. and Nélis, M. (2003a) Surface analysis of mineral substrates for repair works: Roughness evaluation by profilometry and surfometry analysis. *Magazine of Concrete Research*, **55**, 355–366.

Courard, L., Schwall, D., and Piotrowski, T. (2007) Concrete surface roughness characterization by means of opto-morphology technique. In: *Adhesion in Interfaces of Building Materials: A Multi-Scale Approach* (Eds. L. Czarnecki and A. Garbacz), *Advances in Materials Science and Restoration*, AMSR No. 2. Aedificatio Publishers, Freiburg, Germany, pp. 107–115.

Courard, L., Van der Wielen, A., and Darimont, A. (2010) From defects to causes: Pathology of concrete and investigation methods. *Proceedings of the 17th Slovenski kolokvij o betonih*, Ljubljana, Slovenia, May 19, 2010, 10p.

Czarnecki, L., Garbacz, A., and Kurach, J. (2001) On the characterization of polymer concrete fracture surface. *Cement and Concrete Composites*, **32**, 399–409.

Darimont, A., Gilles, P., Dondonné, E., Degeimbre, R., Demars, P., Mertens de Wilmars, A., Lorenzi, G., and Lefèbvre, G. (2006) Re-alkalinization of carbonated concrete by means of alkaline migration. *Bulletin des laboratoires des Ponts et Chaussées*, **261–262**, 121–130.

Dieryck, V., Desmyter, J., Michel, F., and Courard, L. (2005) Surface quality of self-compacting concrete and raw materials properties. In: *Second North American Conference on the Design and Use of Self-Consolidating Concrete (SCC)* and the *Fourth International RILEM Symposium on Self-Compacting Concrete*, Hanley Wood, Addison, IL (Ed. S.P. Shah), Northwestern University, Evanston, IL, pp. 287–295.

Emberson, N.K. and Mays, G.C. (1990) Design of patch repair: Measurements of physical and mechanical properties of repair systems for satisfactory structural performances. In: *Protection of Concrete* (Eds. R.K. Dhir and J.W. Green), E&FN Spon, London, U.K., pp. 867–884.

Emmons, P.H. (1993) *Concrete Repair and Maintenance Illustrated: Problem Analysis, Repair Strategy, Techniques*. R.S. Means Co., Kingston, MA, 295p.

EN 1081: 1998 (1998) Resilient floor coverings. Determination of the electrical resistance. Brussels, Belgium: European Standardization.

EN 12390-3: 2009 (2009) Testing hardened concrete. Compressive strength of test specimens. Brussels, Belgium: European Standardization.

EN 12504-1: 2009 (2009) Testing concrete in structures. Cored specimens. Taking, examining and testing in compression. Brussels, Belgium: European Standardization.

EN 12504-2: 2012 (2012) Testing concrete in structures. Non-destructive testing. Determination of rebound number. Brussels, Belgium: European Standardization.

EN 12504-3: 2005 (2005) Testing concrete in structures. Determination of pull-out force. Brussels, Belgium: European Standardization.

EN 12504-4: 2004 (2004) Testing concrete. Determination of ultrasonic pulse velocity. Brussels, Belgium: European Standardization.

EN 1504-10: 2003 (2003) Products and systems for the protection and repair of concrete structures - Definitions, Requirements, Quality control and evaluation of conformity - Part 10: Site application of products and systems and quality control of the works. Brussels, Belgium: European Standardization.

EN 1766: 2000 (2005) Products and systems for the protection and repair of concrete structures. Test methods. Reference concrete for testing. Brussels, Belgium: European Standardization.

EN 13036-1: 2005 (2005) Road and airfield surface characteristics – Test methods – Part 1: Measurement of pavement surface macrotexture depth using a volumetric patch technique. Brussels, Belgium: European Standardization.

EN 13892-3: 2014 (2014) Methods of test for screed materials. Determination of wear resistance. Böhme. Brussels, Belgium: European Standardization.

EN 13892-4: 2002 (2002) Methods of test for screed materials. Determination of wear resistance-BCA. Brussels, Belgium: European Standardization.

EN 13892-5: 2003 (2003) Methods of test for screed materials. Determination of wear resistance to rolling wheel of screed material for wearing layer. Brussels, Belgium: European Standardization.

EN 14157: 2004 (2004) Natural stones. Determination of abrasion resistance. Brussels, Belgium: European Standardization.

EN ISO 4287: 2009 (2009) Geometrical product specifications (GPS)—Surface texture: profile method—Terms, definitions and surface texture parameters. Brussels, Belgium: European Standardization.

FeBe (1996) *Memento of Architectonic Concrete*, 2nd edn. Bruxelles, Belgium: Belgian Federation of Concrete Industry (in French).

Fukuzawa, K., Mitsui, M., and Numao, T. (2001) Surface roughness indexes for evaluation of bond strengths between CRFP sheet and concrete. In: *10th International Congress on Polymers in Concrete (ICPIC'01)* (Ed. D. Fowler), Honolulu, Hawaii, Paper No. 12.

Garbacz, A., Courard, L., and Bissonnette, B. (2013) A surface engineering approach applicable to concrete repair engineering. *Bulletin of the Polish Academy of Sciences, Technical Sciences*, 61(1), 73–84.

Garbacz, A., Courard, L., and Gorka, M. (2005) Effect of concrete surface treatment on adhesion in repair systems. *Magazine of Concrete Research*, 57, 49–60.

Garbacz, A., Courard, L., and Kostana, K. (2006) Characterization of concrete surface roughness and its relation to adhesion in repair systems. *Materials Characterization*, 56, 281–289.

Garbacz, A. and Garboczi, E. (2003) Ultrasonic evaluation methods applicable to polymer concrete composites, Report NISTIR 6975. United States Department of Commerce, Gaithersburg, MD.

Garbacz, A. and Kostana, K. (2007) Characterization of concrete surface geometry by laser profilometry. In: *Adhesion in Interfaces of Building Materials: A Multi-Scale Approach* (Eds. L. Czarnecki and A. Garbacz), *Advances in Materials Science and Restoration*, AMSR No. 2. Aedificatio Publishers, pp. 147–157.

Gencel, O., Ozel, C., and Filiz, M. (2011) Investigation on abrasive wear of concrete containing hematite. *Indian Journal of Engineering & Materials Sciences*, 18, 40–48.

Gokhale, A.M. and Drury, W.J. (1990) A general method of estimation of fracture surface roughness. Part II. Practical considerations. *Metallurgical Transactions*, 21A, 1201–1207.

Gulikers, J. (2005) Theoretical considerations on the supposed linear relationship between concrete resistivity and corrosion rate of steel reinforcement. *Materials and Corrosion*, 56(6), 393–403.

Horszczaruk, E. (2004) The model of abrasive wear of concrete in hydraulic structures. *Wear*, 256, 787–796.

Horszczaruk, E. (2008) *The Model of Abrasive Wear of Cement Concrete*. Szczecin, Poland: Publishing House of the West-Pomeranian University of Technology.

ICRI Guideline No. 03732 (2002) Selecting and specifying concrete surface preparation for sealers, coatings, and polymer overlays, Rosemont, IL: International Concrete Repair Institute.

Jasim, A.T. and Jawad, F.A. (2010) Effect of oil on strength of normal and high performance concrete. *Al-Qadisiya Journal for Engineering Sciences*, 3(1), 24–32.

Justnes, H. (1995) Capillary suction of water by polymer cement mortars. In: *Proceedings of the RILEM Symposium on Properties and Test Methods for Concrete-Polymer Composites (ICPIC)*, Oostende, Belgium (Ed. D. Van Gemert), pp. 29–37.

Kurzydłowski, K.J. and Ralph, B. (1996) *Quantitative Description of Microstructure*. New York: CRC Press.

Lallemant, I., Rougeau, P., Gallias, J.L., and Cabrillac, R. (2000) Contribution of microscopy to the characterization of concrete surfaces presenting local tint defects. *22nd International Conference on Cement Microscopy*, April 29–May 4, Montréal, Quebec, Canada, pp. 107–121.

Laplante, P., Aïtcin, P.C., and Vezina, D. (1991) Abrasion resistance of concrete. *Journal of Materials in Civil Engineering*, 3(1), 19–28.

Lemaire, G. (2003) Contribution to the quality control and assessment of concrete surfaces. PhD Thesis, Université Paul Sabatier, Toulouse, France, 209p. (in French).

Lemaire, G., Escadeillas, G., and Ringot, E. (2005) Evaluating concrete surfaces using an image analysis process. *Construction and Building Materials*, 19, 604–611.

Kanare, H.M. (2005) *Concrete Floors and Moisture*. Skokie, IL: Portland Cement Association, Engineering Bulletin, 119, p. 156.

Maage, M. (2004) The new European EN 1504 standard. Guidelines for consultant. *NORECON Seminar*, Kopenhagen, Denmark.

Maerz, H., Chepur, P., Myers, J., and Linz, J. (2001) Concrete roughness characterization using laser profilometry for fiber-reinforced polymer sheet application. *80th Annual Meeting*, Transportation Research Board, Washington, DC, Paper No. 01-0139.

Malhotra, V.H. (1977) Concrete strength requirements: Core vs. in situ evaluation. *Proceedings of the American Concrete Institute*, 74(4), 163–172.

Martin, M. (2007) Study of the texture of vertical formworked concrete surfaces. PhD Thesis, Université Laval, Quebec, Canada and Université de Cergy-Pontoise, Cergy-Pontoise, France, 225p (in French).

Martin, M., Gilson, A., and Courard, L. (2008) Aesthetic evaluation and control of concrete building surfaces. *Materials and Sensations*, October 22–24, Pau, France, 4p.

Morris, J., Moreno, E.I., and Sagues, A.A. (1996) Practical evaluation of resistivity of concrete in test cylinders using a Wenner array probe. *Cement and Concrete Research*, 26(12), 1779–1787.

Naderi, M., Cleland, D.J., and Long, A.E. (1986) In situ test methods for repaired concrete structures. In: *Proceedings of the International Symposium on Adhesion between Polymers and Concrete*, Aix-en-Provence, France (Ed. H.R. Sasse), Chapman & Hall, London, U.K., pp. 707–718.

Neville, A.M. (2012) *Properties of Concrete*. Pearson Education Limited, Harlow, UK, 846pp.

Neville, A.M. (2006) *Concrete: Neville's Insights and Issues*. Thomas Telford Publishing, London, U.K., 320p.

NIT 216 (2000) Industrial floors with reactive resinous products. Belgian Building Research Center, Brussels, Belgium (in French).

NT Build 439 (November 1995). Concrete, hardened: Relative humidity measured in drilled holes. Espoo, Finland: Nordtest.

Perez, F., Bissonnette, B., and Courard, L. (2009) Combination of mechanical and optical profilometry techniques for concrete surface roughness characterization. *Magazine of Concrete Research*, **61**(6), 389–400.

Pleau, R., Henné, J., Boulet, D., Blais, D., and St-Pierre, M. (1999) Aesthetic properties of architectonic high-performance self-levelling concretes. *Proceedings of the International Conference Creating with Concrete, Radical Design and Concrete Practices*, Dundee, Scotland, pp. 191–200.

Ramezanianpour, A.A., Pilvar, A., Mahdikhani, M., and Moodi, F. (2011) Practical evaluation of relationship between concrete resistivity, water penetration, rapid chloride penetration and compressive strength. *Construction and Building Materials*, **25**, 2472–2479.

RILEM TC 25-PEM (1980) Recommendation tests to measure the deterioration of stones and to assess the effectiveness of treatment methods. *Materials and Structures*, **13**(75), 175–254.

Robery, P.C. (1995) Investigation methods utilizing combined NDT techniques. *Construction Repair*, **9**(6), 11–16.

Santos, P. and Júlio, E. (2008) Development of a laser roughness analyzer to predict in-situ the bond strength of concrete-to-concrete interfaces. *Magazine of Concrete Research*, **60**(5), 329–337.

Santos, P. and Júlio, E. (2010) Comparison of methods for texture assessment of concrete surfaces. *ACI Materials Journal*, **107**(5): 434–440.

Santos, P. and Júlio, E. (2013) State-of-the-review on roughness quantification methods for concrete surfaces. *Construction and Building Materials*, **38**(5): 912–923.

Saouma, V.E. and Barton, C.C. (1994) Fractals, fractures, and size effects in concrete. *Journal of Engineering Mechanics*, **120**, 835–854.

Sherrington, I. and Smith E.H. (1988) Modern measurement techniques in surface metrology—Part I: Stylus instruments, electron microscopy and non-optical comparators. *Wear*, **125**, 271–288.

Siewczyńska, M. (2008) Effect of selected properties of concrete on adhesion of protective coating, Ph.D. Thesis, Poznań University of Technology, Poznań, Poland.

Silfwerbrand, J. (1990) Improving concrete bond in repaired bridge decks. *Concrete International*, **12**, 61–66.

STP 169C (1994) *Significance of Tests and Properties of Concrete and Concrete-Making Materials, Part 3* (Eds. P. Klieger and J.F. Lamond), ASTM, Philadelphia, PA, 239p.

Stroeven, P. (2000) A stereological approach to roughness of fracture surface and tortuosity of transport paths in concrete. *Cement and Concrete Composites*, **22**, 331–341.

Teodoru, G.Y.M. and Mommens, A. (1991) Non-destructive testing in the quality control of buildings: Why, what and how? In: *Proceedings of the Symposium on Quality Control of Concrete Structures, Gent* (Eds. L. Taerwe and H. Lambotte), E &FN Spon, London, U.K., pp. 367–376.

Thompson, M.S. (1969) Blowholes in concrete surfaces. *Concrete*, **3**(2), 64–66.

Underwood, E.E. (1987) Stereological analysis of fracture roughness parameters. *Acta Stereologica*, **6**, 170–178.

Van der Wielen, A., Courard, L., and Nguyen, F. (2012) Static detection of thin layers into concrete with ground penetrating radar. *Restoration of Buildings and Monuments*, **18**(3/4), 247–254.

Wojnar, L. (1995) *Image Analysis: Applications in Materials Engineering.* New York: CRC Press.

Wong, H.S. and Buenfeld, N.R. (2006) Patch microstructure in cement-based materials: Fact or artefact? *Cement and Concrete Research*, **36**, 990–997.

Wilson, S.A., Langdon, N.J., and Walden, P.J. (July 1, 2001) The effects of hydrocarbon contamination on concrete strength. *Proceedings of the ICE—Geotechnical Engineering*, **149**(3), 189 –193.

Zajc, A., Courard, L., Garbacz, A., and Wolff, L. (2006) Testing procedures and other regulations (Chapter 6). RILEM TC 184—IFE Industrial floors for withstanding environmental attacks, including repair and maintenance. RILEM Report 33 (Ed. P. Seidler), RILEM Publications S.A.R.L, Bagneux, France, pp. 101–109.

Chapter 4

Interfacial phenomena

4.1 ADHESION: PRINCIPLES

Adhesion has already been defined by means of *specific* and *mechanical* considerations (see Figure 1.1). Many authors studied the mechanism of adhesion and developed theories based on the physicochemical interactions between solids and liquids (Figure 4.1). Each individual model of adhesion can contribute to an explanation of the phenomenon but is unable, by itself, to give a comprehensive explanation (Fiebrich, 1994).

If there is creation of an interface, it means that there is a potential and mutual interaction between two bodies. A physiological transposition of the situation would let us say that it is a case of appetency (from Latin word *"appetentia"*): this term describes the natural attraction towards something in order to satisfy a need. Its application to the interface between concrete and a new layer summarizes all the parameters—physical, chemical, and mechanical properties—of the substrate, the new layer, and the environment that influence the creation and the stability of this interface. Finally, it will lead to the macroscopical effect and sign of the efficiency of the cohesion: bond strength (Courard and Garbacz, 2010). One interesting approach is based on the analysis of the thermodynamic equilibrium of a solid/liquid.

4.2 ADHESION: THERMODYNAMIC APPROACH

Adhesion represents the bond created between two different bodies and is thus a surface problem. The equilibrium between a liquid and a solid surface is described by the following equation proposed by Young and Dupré and illustrated in Figure 4.2:

$$\gamma_{SV} = \gamma_{SL} + \gamma_L \cos \theta \tag{4.1}$$

where γ_{SV}, γ_{SL}, and γ_L are the surface free energy of the solid in contact with vapor, the interfacial free energy, and the surface free energy of the liquid in contact with vapor, respectively.

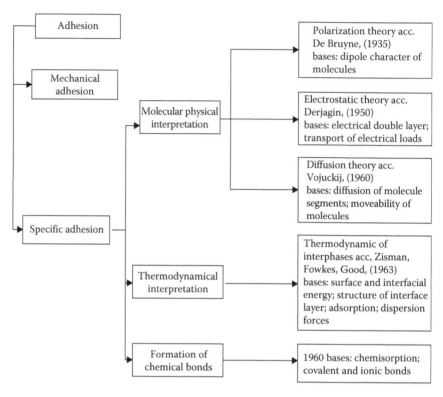

Figure 4.1 Theories of adhesion. (From Fiebrich, M.H., Scientific aspects of adhesion phenomena in the interface mineral substrate-polymers, in: Wittman, F.H., ed., *Proceedings of the Second Bolomey Workshop on Adherence of Young and Old Concrete*, Aedificatio Verlag, Unterengstringen, Switzerland, 1994, pp. 25–58.)

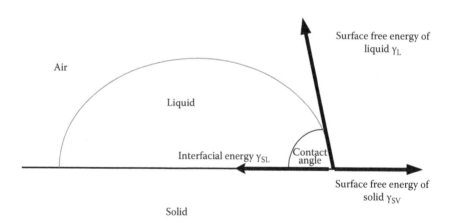

Figure 4.2 Wettability of a solid surface by a liquid in accordance with the Young–Dupré equation. (From the authors.)

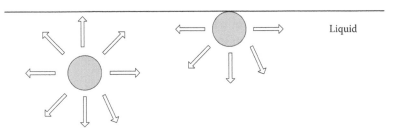

Figure 4.3 Attraction forces between molecules located at the surface and inside the material. (From the authors.)

Surface free energy is a direct measurement of intermolecular forces (Courard, 2002). The free energy in the surface layer is the result of the attraction of the bulk material for the surface layer, and this attraction tends to reduce the number of molecules in the surface layer, resulting in an increased intermolecular distance (Figure 4.3). In order to re-establish equilibrium, a work must be produced: this explains why free energy exists and why there is a surface free energy. Surface energy quantifies the disruption of intermolecular bonds that occur when a surface is created (Derjagin, 1978). Surface energy may therefore be defined as the excess energy at the surface of a material compared to the bulk.

The most common types of physical attractive forces are Van der Waal's forces, and they can be attributed to different causes (Courard, 2000):

- Dispersion forces arising from internal electron motions that are independent of dipole moments
- Polar forces arising from the orientation of permanent electric dipoles and the induction effect of permanent dipoles on polarizable molecules

The determination of the contact angle between a liquid and the solid surface is the most widely used technique to characterize surface properties of solids. Contact angle is often used as a measure of surface hydrophobicity: the higher the contact angle, the more hydrophobic the solid surface.

Based on the thermodynamic equilibrium between a liquid drop and the solid surface, the analysis and the determination of the contact angle and the surface free energy of liquids give us the respective effects of polar and dispersion forces.

4.3 CONTACT ANGLE AND INTERFACIAL FREE ENERGY

4.3.1 Introduction

The surface free energy of a substrate determines its ability to be wetted by a liquid, which contributes to increasing the contact surface and, consequently, the effective adhesion forces. The second law of thermody-namics indicates that a system with two phases is stable if the local energy is minimum. In the case of an interface between concrete and a new layer of concrete, repair material, or coating, it is assumed that minimum inter-facial free energy will yield

- Maximum bond strength
- Maximum durability of the bond

It is important to make a distinction between interface bond *development* and *stability*. There is of course no stability without creation, and the two concepts are fundamentally linked. Concrete surface engineering primar-ily deals with all the phenomena and properties involved at the interface, which rule the initiation and development of bond as well as its stability.

Thermodynamics provides a basis for a physical description of the equi-librium at the interface, which is the fundamental condition to develop a bond. Nevertheless, it is not as such a sufficient condition for a strong and durable bond. Getting an insight of surface free energy of the solid sub-strate will help to better understand conditions of creation and development of adhesion.

4.3.2 Fundamentals and spreading conditions

Analytical parameters are defined hereafter in order to express the equilib-rium at the interface between a solid substrate and a liquid layer. Work of adhesion (W_A) corresponds to the energy needed to separate the two layers to be bonded and is determined as follows:

$$W_A = \gamma_L + \gamma_{SV} - \gamma_{SL} \tag{4.2}$$

Taking into account the Young–Dupré relationship (Equation 4.1), one obtains

$$W_A = \gamma_L(1 + \cos\theta)$$

The assumption that the work of adhesion is greater than the work of cohe-sion, which would lead to spreading, can be described as follows:

$$W_A > W_C \tag{4.3}$$

where $W_C = 2\gamma_L$.

In such a condition, it is possible to define the *spreading coefficient* as

$$S = W_A - W_C \geq 0 = \gamma_{SV} + \gamma_L - \gamma_{SL} - 2\gamma_L \geq 0 = \gamma_{SV} - \gamma_L - \gamma_{SL} \geq 0 \tag{4.4}$$

Using the Young–Dupré relationship (Equation 4.1), one obtains

$$S = \gamma_L(\cos \theta - 1) \geq 0 \qquad (4.5)$$

If S has a negative value, spreading is not complete and the contact angle is given by Equation 4.1 (Young–Dupré). If S has a positive value, it means that the equation has no solution and that there is complete spreading of the liquid onto the solid surface, ending up in $\theta = 0$.

4.3.3 Definitions of interfacial parameters

Zisman and Fox (1950, cited by Kinloch (1987)) observed experimentally that for most of the substrates, there is a linear relationship between $\cos \theta$ and γ_L:

$$\cos\theta = \begin{cases} (1+m) - m\dfrac{\gamma_L}{\gamma_C} & \text{for } \gamma_L > \gamma_C \\ 1 & \text{for } \gamma_L \leq \gamma_C \end{cases} \qquad (4.6)$$

where γ_C is the critical surface free energy of the solid and m is the Y-intercept on the graph of Figure 4.4. Gutowski (1985a,b, 1987) has also shown that the assumption of a linear relation is valid for $0.6 < \cos \theta < 1$.

Using the Young–Dupré equation (Equation 4.1), one derives

$$\gamma_{SL} = \begin{cases} \gamma_S - \gamma_L(1+m) + m\dfrac{\gamma_L^2}{\gamma_C} & \text{for } \gamma_L > \gamma_C \\ \gamma_S - \gamma_L & \text{for } \gamma_L \leq \gamma_C \end{cases} \qquad (4.7)$$

The minimum interfacial free energy may be calculated by differentiating Equation 4.7 with respect to γ_L:

$$\frac{\partial \gamma_{SL}}{\partial \gamma_L} = 0 \quad \text{and} \quad \frac{\partial^2 \gamma_{SL}}{\partial \gamma_L^2} > 0 \qquad (4.8)$$

which yields (for $m \leq 1$)

$$\gamma_L = \frac{\gamma_C(1+m)}{2m} \qquad (4.9)$$

$$\gamma_{SL}^{min} = \gamma_S - \frac{\gamma_C(1+m)^2}{4m} \qquad (4.10)$$

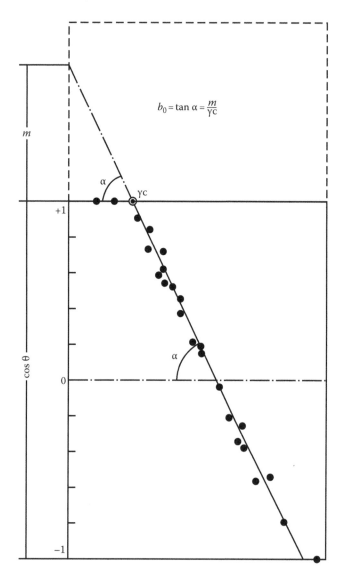

Figure 4.4 Zisman's linear relationship describing the wettability of a solid (where the term *m* corresponds to the Y-coordinate intercepted by the regression curve). (From Courard, L. et al., *Constr. Build. Mater.*, 25(1), 260, 2011.)

If $m > 1$, from the definition of γ_{SL}, one gets

$$\gamma_{SL}^{min} = \gamma_S - \gamma_C \quad \text{for } \gamma_L = \gamma_C \tag{4.11}$$

Dimensionless parameters are typically used for characterizing adhesion (Gutowski, 1985b):

$$\left[\frac{W_A}{\gamma_L}\right] \rightarrow \text{relative work of adhesion} \tag{4.12}$$

$$\left[\frac{\gamma_{SL}}{\gamma_L}\right] \rightarrow \text{relative interfacial free energy} \tag{4.13}$$

$$\left[\frac{\gamma_S}{\gamma_L}\right] \rightarrow \text{modulus of energy of the system} \tag{4.14}$$

Equation 4.7 may thus be rewritten as follows:

$$\frac{\gamma_{SL}}{\gamma_S} = \begin{cases} 1 - \dfrac{\gamma_L}{\gamma_S}(1+m) + m\dfrac{\gamma_L^2}{\gamma_S \cdot \gamma_C} & \text{for } \gamma_L > \gamma_C \\[3mm] 1 - \dfrac{\gamma_L}{\gamma_S} & \text{for } \gamma_L \leq \gamma_C \end{cases} \tag{4.15}$$

$$\frac{\gamma_{SL}}{\gamma_L} = \begin{cases} \dfrac{\gamma_S}{\gamma_L} + m\dfrac{\gamma_L}{\gamma_C} - (1+m) & \text{for } \gamma_L > \gamma_C \\[3mm] \dfrac{\gamma_S}{\gamma_L} - 1 & \text{for } \gamma_L \leq \gamma_C \end{cases} \tag{4.16}$$

Based upon Zisman's equation (Equation 4.6), spreading occurs when the γ_{SL}/γ_L ratio reaches a threshold value given by the following expression:

$$\frac{\gamma_S}{\gamma_L} = \frac{1}{\phi_0^2} \tag{4.17}$$

where ϕ_0 is the effective bonding factor defined by Good and Girofalco (Gutowski, 1985a) and is constant for a given solid.

Substituting γ_S/γ_L with a coefficient (a_S) and rearranging the terms, Equation 4.7 may be rewritten as follows:

$$\frac{\gamma_{SL}}{\gamma_S} = \begin{cases} 1 + \dfrac{ma_S}{a^2} - \dfrac{1+m}{a} & \text{for } 0 < a < a_S \\[3mm] 1 - \dfrac{1}{a} & \text{for } a \geq a_S \end{cases} \tag{4.18}$$

or

$$\frac{\gamma_{SL}}{\gamma_L} = \begin{cases} a - (1+m) - \dfrac{ma_S}{a} & \text{for } 0 < a < a_S \\ a - 1 & \text{for } a \geq a_S \end{cases} \qquad (4.19)$$

Expected bond performance is related to minimization of the relative interfacial free energy, which happens when

$$a_{min} = \left(\frac{\gamma_S}{\gamma_L}\right)_{min} = \begin{cases} \dfrac{2m}{(1+m)\phi_0^2} & \text{for } m \leq 1 \\ \dfrac{1}{\phi_0^2} & \text{for } m > 1 \end{cases} \qquad (4.20)$$

As already explained, performance analysis of a bond is mainly based on the spreading ability of a given liquid when in contact with a given solid surface. The next procedure may be selected in order to solve the analytical system:

1. Modulus of energy of the system

$$a = \frac{\gamma_S}{\gamma_L} \qquad (4.21)$$

2. Effective bonding factor

$$\phi_0 = \sqrt{\frac{\gamma_C}{\gamma_S}} \qquad (4.22)$$

3. Threshold modulus of energy for perfect spreading

$$a_S = \frac{1}{\phi_0^2} = \frac{\gamma_S}{\gamma_C} \qquad (4.23)$$

4. Relative interfacial energy

$$\frac{\gamma_{SL}}{\gamma_S} \qquad (4.24)$$

5. Modulus of energy corresponding to γ_{SL}

If substrate spreading conditions are unknown, it can be assumed in a first-order analysis (Gutowski, 1987) that $m = 1$ and $\gamma_S = \gamma_C$ and, as a result, that γ_{SL} is minimized (for $a_{min} = 1$ and the spreading threshold value $a_S = 1$). The relationships are simplified as follows:

$$\frac{\gamma_{SL}}{\gamma_S} = \begin{cases} \left(1 - \dfrac{1}{a}\right)^2 = \left(1 - \dfrac{\gamma_L}{\gamma_S}\right)^2 \\[2ex] \left(1 - \dfrac{1}{a}\right) = \left(1 - \dfrac{\gamma_L}{\gamma_S}\right) \end{cases} \tag{4.25}$$

and $\gamma_{SL}^{min} = 0$ for $a_{min} = 1$.

An important remark needs to be made with regard to the material changes occurring after placement and their incidence upon surface free energy. Some authors (Kinloch, 1987) have made the assumption that surface free energy does not change with the change from liquid to solid during polymerization or hydration of the repair or surface treatment material. Replacing γ_L with γ_L^C (surface free energy of the polymerized liquid), adhesives in their solid state are considered able to "fictively" spread over the solid substrate surface, the wetting process being characterized by an "imaginary" contact angle.

Some authors (Gutowski, 1987; Kinloch, 1987) in fact postulated that $\gamma_L^C = \gamma_L$, and for $\gamma_L^C = \gamma_C$, γ_{SL} is a linear function of γ_L^C, while for $\gamma_L^C = \gamma_C$, it is equal to $\gamma_S - \gamma_L^C$.

4.3.4 Relationships between adhesion and thermodynamic parameters

What really matters for the materials engineer is finally to yield, for a given combination of materials, the optimum bond strength, both in tension and shear. What can be the relationship between the bond and the main parameters that were just described?

Interesting relationships can be established between the strength of a system and both its modulus of energy and relative interfacial energy (Gutowski, 1987):

$$\tau = f\left(a = \frac{\gamma_S}{\gamma_L^C}\right) \tag{4.26}$$

$$\tau = f\left(\frac{\gamma_{SL}}{\gamma_L^C}\right) \tag{4.27}$$

This type of relationship has been established for metallic substrates (Figure 4.5), where it clearly appears that the maximum bonding strength is achieved for $a = \gamma_S/\gamma_L^C = 1$. Above or beneath this value, strength dramatically decreases (at a faster rate for $a < 1$ than $a > 1$).

The relationship between interfacial energy and relative adhesion is well illustrated in tests (Gutowski, 1987) using a "fiberglass/epoxy resin" composite, with the glass fibers having undergone various surface treatments (rinsing with water, silanes, silicones, or polyvinyl acetate).

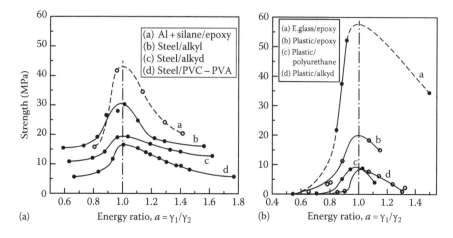

Figure 4.5 Relationship between bond strength and modulus of energy for composite sytems (a) with a metallic substrate and (b) with a nonmetallic substrate. (From Gutowski, W., *Int. J. Adhes. Adhesiv.*, 7(4), 189, 1987.)

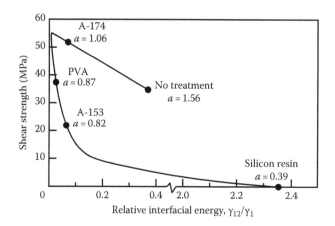

Figure 4.6 Relationship between shear bond strength and relative interfacial energy $(\tau = f(\gamma_{SL}/\gamma_S))$ for a glass E/polyester resin composite following different fiber surface treatments. (From Gutowski, W., *Int. J. Adhes. Adhesiv.*, 7(4), 189, 1987.)

Treatment of the fibers gives rise to variable results, mainly related to the value of parameter *a* (Figure 4.6). The thermodynamic condition for minimum interfacial energy explains the two types of curves found for the relationship $\tau = f(\gamma_{SL}/\gamma_S)$ of different systems:

- A lower curve, almost hyperbolic, for systems with a modulus of energy lower than a_{min}
- An upper curve for systems with a modulus of energy greater than a_{min}

In summary, from a thermodynamic point of view, the mechanical strength at the interface between two materials bonded to each other is ruled by the following function:

$$\tau = f\left[\gamma_{SL} = f_2(\gamma_S, \gamma_L, \phi_0, m)\right] \tag{4.28}$$

Nevertheless, knowledge of the interfacial surface tension and its minimization is not sufficient to predict the bond strength between two materials. The modulus of energy also needs to be assessed as for a given value of γ_{SL}, γ_{SL}/γ_S, or γ_{SL}/γ_L, different bond strength values may be yielded, as shown in Figure 4.5.

4.4 EVALUATION OF INTERFACIAL ENERGIES

Interfacial energy is a residue of the initial surface energy balance that was not equilibrated. This interfacial energy (Figure 4.2), as well as the respective liquid and solid surface energies, can be described with two terms relative to polar (γ^p) and dispersive (γ^d) interactions (Comyn, 1992; Comyn et al., 1993).

The surface free energy of the liquid can be evaluated with regard to the contribution of dispersive and polar forces of the solid. Owens and Wendt (Courard, 2002) defined γ_{SL} as the geometric mean between these two effects:

$$\gamma_{SL} = \gamma_S + \gamma_L - 2\left(\gamma_S^d \cdot \gamma_L^d\right)^{1/2} - 2\left(\gamma_S^p \cdot \gamma_L^p\right)^{1/2} \tag{4.29}$$

Combining the latter expression with the Young–Dupré equation (Equation 4.1) and using reference liquid and solid combinations with known contact angle values (Equation 4.30), the polar and dispersive components of a solid can be determined as follows:

$$1 + \cos\theta = \frac{2}{\gamma_L}\left[\left(\gamma_S^d \cdot \gamma_L^d\right)^{1/2} + \left(\gamma_S^p \cdot \gamma_L^p\right)^{1/2}\right] \tag{4.30}$$

To evaluate the interactions at the interface, it is necessary to determine parameters γ_L, γ_L^p, γ_L^d and γ_S, γ_S^p, γ_S^d, and γ_{SL}.

The surface free energy of the liquid phase is evaluated using the Wilhelmy scale (Courard, 2002), while contact angles are measured with a goniometer. The process by which the dispersion and polar components of γ_L and γ_S can be determined is as follows (Courard et al., 2011):

1. Consider the surface free energy of a selected liquid characterized by dispersion forces (γ_L^d) only.

2. Select reference solids with surface free energy characterized by dispersion forces (γ_S^d) only.
3. By measuring the contact angles of well-characterized liquids (known value of γ_L) on reference solids, determine the dispersion and, by subtraction, the polar component of the surface energy of these liquids.
4. By measuring contact angles of liquids on solids, evaluate the two components of γ_S.

The contact angle measurements are easy to perform on a smooth and flat surface; however, numerous parameters can actually affect the measurement reliability, particularly in the case of cementitious materials:

- The surface *roughness* (from 1 to 1000 μm) scale is largely beyond the size of the molecules. Wenzel (cited in Courard et al., 2011) tried to quantify this effect by means of a roughness factor r_f calculated as follows:

$$r_f = \frac{\text{real (specific) surface}}{\text{geometric surface}} \tag{4.31}$$

which is introduced in the Young–Dupré equation in the following form:

$$\cos\theta_f = r_f\cos\theta_S \tag{4.32}$$

where θ_f and θ_S represent contact angles measured on rough and flat surfaces, respectively. If θ_S is smaller than 90° on a flat surface, any increase in roughness (and consequently, in the value of r_f) will result in a decrease of θ_f, thus an increase in apparent superficial tension of the solid phase and, consequently, a greater wettability. However, if, for a flat surface, θ_S is higher than 90°, a rougher surface will automatically induce an increase of θ_f.

- Surface cavities and pores can have both positive and negative effects. Large cavities in combination with highly viscous fluids can promote the entrapment of air bubbles at the interface, which in fact is one of the most common technological causes for defects of bonds and loss of adhesion. The influence of air inclusion upon solid surface wettability is described by the following equation:

$$\cos\theta_{ab} = A_a \cdot \cos\theta_a + A_b \cdot \cos\theta_b \tag{4.33}$$

This is valid for a two-phase (a and b) solid surface, with occupied surfaces A_a and A_b, respectively. If phase b represents pores or cavities filled with air, the contact angle increases to 180° (because there is no wetting) and Equation 4.33 becomes:

$$\cos\theta_{ab} = A_a \cdot \cos\theta_a - A_b \tag{4.34}$$

The contact angle θ_{ab} decreases only if the liquid is able to penetrate into the capillaries. In this regard, a liquid of low viscosity spreads more easily into surface cavities and fill porosity, thus increasing the effective surface and promoting stronger adhesion.

• Surface energy discrepancies (differences of polarities due to differences in molecules or atoms present at the surface) can also promote dispersion of the value of the angle of contact through a modification of the liquid bead shape; it is then like a submicroscopic roughness.

4.5 EXPERIMENTAL STUDIES

4.5.1 Preparation of repair materials

It is not practically possible to determine surface free energy of liquids with high viscosity. The investigation on mortars, slurries, and other thick cement-based "liquids" is thus complex as shear and viscosity forces inside the material counteract the measurement of the surface properties.

In a study by Courard (2002), the yield stress (τ_m) and viscosity (η) of slurries used as bonding agents between ordinary concrete supports and different (cement concrete or portland cement concrete) repair mortars were reduced with selected admixtures. The following procedure was developed for obtaining liquid slurries that would allow surface energy determination of both the liquid and the solid phases:

• Mixing water and cement CEM I 42.5 (W/C = 0.5) for 3 min and pouring in the centrifugation device containers
• Centrifuging for 20 min at 6000 rpm
• Collecting the centrifuged solutions (±100 mL) into sealed containers after filtration with *Whatman*® paper no. 51 (retention of 20–25 μm particles)

It is essentially this latter portion of the slurries that can actually penetrate into the superficial porosity of concrete, considering the relative dimension of cement particles and the accessible pore radius. In view of analyzing the thermodynamic equilibrium at the interface and wetting of concrete by cement slurries (modified or not), the use of centrifuged solutions is more representative of real phenomena, while preventing the interactive effect of viscosity on contact angle measurement.

4.5.2 Surface free energy of liquids

The determination of the surface energy of liquids can be performed using the *Wilhelmy* plate (Figure 4.7) or the *de Nouÿi* ring (Figure 4.8) and a

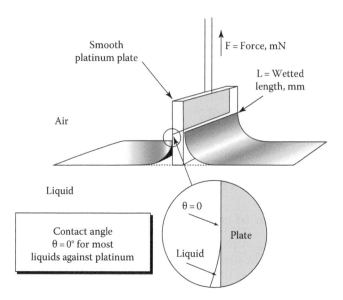

Figure 4.7 Wilhelmy superficial tension measurement system. (From Michel, F. et al., *Cement Concr. Compos.*, 45, 111, 2014.)

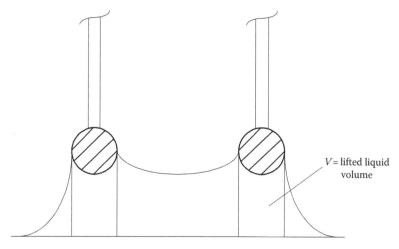

Figure 4.8 De Nouyï superficial tension measurement system. (From Michel, F. et al., *Cement Concr. Compos.*, 45, 111, 2014.)

tensiometer, based on the measurement of the weight necessary to pull out the metallic object from the liquid (Herkins and Jordan, 1930; Possert and Kamusewitz, 1993).

Table 4.1 presents the results of surface free energy evaluation of distilled water and centrifuged solutions of cement slurries. Only a solution with vinyl copolymer seems to induce a significant modification of surface free energy.

Table 4.1 Surface free energy of centrifuged solutions of cement slurries modified with admixtures and reference liquids

Reference	Temperature (°C)	Surface free energy (mN/m)
Distilled water	23.2	71.1
Melamine (macromolecules)	23.5	66.3
Melamine	23.3	70.3
Naphtalene	23.2	67.3
Vinyl copolymer	23.3	49.1
Maleic acid	23.2	67.8
Natrium lignosulfonate	23.3	66.3
Cement-based slurry (no admixture)	23.3	70.6
Dimethylformamide	23.1	36.3
2-Dimethylethanolamine	22.9	27.95
3-Dimethylamino-1,2-propanediol	22.8	36.1
Tetramethylene sulfone	23	49.6
α-Bromonaphtalene	23.1	42.5

Four different products (pure chemical products) are used as reference liquids, while α-bromonaphtalene is a chemical substance where only dispersion forces are able to interact ($\gamma_L = \gamma_L^d$). This liquid allows to determine the dispersion component of the solid surface.

4.5.3 Surface free energy of solids

As already explained (§4.4), surface free energy of solids can be evaluated by determining the value of the contact angle of reference liquids (with well-known and nonvariable surface energies) with a goniometer (Figure 4.9). The results of such characterization experiments were reported by Courard (2002).

Due to the heterogeneity of concrete, the measurement was separately performed on the mortar and the aggregate. The influence of the mineral nature (Figure 4.10) of the aggregates had been investigated in a previous study (Sasse and Fiebrich, 1983).

Three main supports were tested:

- Cement paste W/C = 0.4
- Cement paste W/C = 0.4 + hydrophobic treatment
- Natural limestone

Hydrophobic treatment was carried out by the application of DYNASYLAN BSM 100W (≈144 g/m²). Evolution of contact angle versus time was taken into account by measuring the contact angle at different time steps.

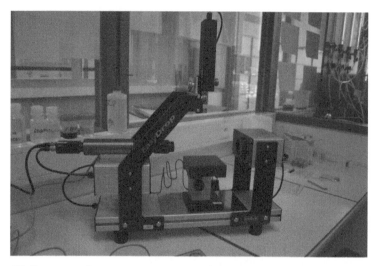

Figure 4.9 Contact angle measurement system. (From the authors.)

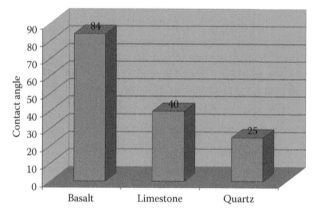

Figure 4.10 Contact angle of epoxy resin on different types of aggregates. (From Fiebrich, M.H., Scientific aspects of adhesion phenomena in the interface mineral substrate-polymers, in: Wittman, F.H., ed., *Proceedings of the Second Bolomey Workshop on Adherence of Young and Old Concrete*, Aedificatio Verlag, Unterengstringen, Switzerland, 1994, pp. 25–58.)

Table 4.2 presents the results of the contact angle measurements. For the contact phenomenon study, instantaneous contact angle data are acceptable because it is not significantly influenced by capillary absorption or roughness (Courard, 1999). However, if the intent is to investigate the suction properties, contact angle data should be recorded for at least 1 or 2 minutes.

A statistical treatment, based on multivaried variance analysis, was carried out with contact angle as a dependent variable and time, type

Table 4.2 Evolution of contact angle on a cement paste and limestone rock for reference liquids, water, and centrifuged solutions of cement slurries modified with different admixtures versus time

Reference	Cement paste		Cement paste + hydrophobic treatment		Limestone	
	20 s	2 min	20 s	2 min	20 s	2 min
Melamine (macromolecules)	12	9	115	113	16	11
Melamine	17	14	126	123	19	13
Naphtalene	19	15	130	129	23	17
Vinyl copolymer	21	18	122	120	21	16
Maleic acid	32	29	111	110	23	18
Natrium lignosulfonate	23	21	127	125	29	20
Cement-based slurry (no admixture)	34	31	124	120	21	26
Distilled water	58	50	120	116	43	35
Dimethylformamide	22	—	49	—	0	—
2-Dimethylethanolamine	22	—	0	—	0	—
Tetramethylene sulfone	59	—	112	—	21	—
α-Bromonaphtalene	11	—	66	—	0	—

Source: Courard, L., *Mater. Struct.*, 35, 149, 2002.

of support, and type of liquid as independent variables. The conclusions drawn by Courard (2002) were as follows:

- *Time* has a significant effect on contact angle values.
- *Support* has a significant effect on contact angle values; it is quite significant and evident for support with hydrophobic treatment.
- *Reference liquid* has a significant effect on contact angle values.

Other authors (e.g., Kinloch, 1987; Possart and Kamusewitz, 1992) had reported this time–temperature dependence.

4.5.4 Determination of dispersion and polar components of liquids and solids

Parafilm M® is a plastic polyethylene film (Rouxhet, 1996) characterized by a polar component of its surface free energy equal to zero. If we use α-bromonaphtalene ($\gamma_S^p = 0$), we will be able to deduce, from Equation 4.30, the value of $\gamma_S^d = 0$ (Equation 4.35):

$$\gamma_S^d = \frac{(\cos\theta_{brom} + 1)^2}{4}\gamma_{brom} \tag{4.35}$$

Table 4.3 Contact angles on reference solid and dispersion and polar component of liquids

Reference	Contact angle on parafilm (°)	Dispersion component, γ_L^d (mN/m)	Surface free energy, γ_L (mN/m)	Polar component, γ_L^p (mN/m)
Distilled water	101	21.7	71.1	49.4
Dimethylformamide	52	22.5	36.3	13.8
Tetramethylene sulfone	71	28.3	49.6	21.3
2-Dimethylethanolamine	35	16.9	27.95	11.05
3-Dimethylamino-1,2-propanediol	64	17.7	36.1	18.4
Melamine (macromolecules)	98	21.35	66.3	44.95
Melamine	100	22.1	70.3	48.2
Naphtalene	94	25.7	67.3	41.6
Vinyl copolymer	94	13.7	49.1	35.4
Maleic acid	99	21.4	67.8	46.4
Natrium lignosulfonate	100	19.7	66.3	46.6
No admixture	93	29.3	70.6	41.3

Source: Courard, L., *Mater. Struct.*, 35, 149, 2002.

Measurements give a value of $\gamma_S^d = 38.15$ mN/m, which is in accordance with literature (Rouxhet, 1996). The determination of contact angles of different liquids with parafilm (Table 4.3) allows to calculate the dispersion component of the surface free energy of these liquids because there is no possible polar interaction. By subtracting this value to γ_L previously determined, it is possible to evaluate γ_L^p for each liquid. The calculation of the dispersion component of liquid comes from Equation 4.30, where $\gamma_S^p = 0$ (Equation 4.36):

$$\gamma_L^d = \frac{1}{\gamma_S^d}\left(\frac{(1+\cos\theta)}{2}\cdot\gamma_L\right)^2 \qquad (4.36)$$

Values obtained for distilled water are in accordance with literature (Kamusewitz and Possart, 1985; Comyn, 1992). A similar development can be set up to calculate dispersion and polar components of solids. From the equation of Owens and Wendt (Equation 4.30), it is possible to determine, by linear regression, the rate and the Y-axis intercept value for the following equation:

$$\frac{\gamma_L(1+\cos\theta)}{2\left(\gamma_L^d\right)^{1/2}} = \left(\gamma_S^p\right)^{1/2}\cdot\frac{\left(\gamma_L^p\right)^{1/2}}{\left(\gamma_L^d\right)^{1/2}} + \left(\gamma_S^d\right)^{1/2} \qquad (4.37)$$

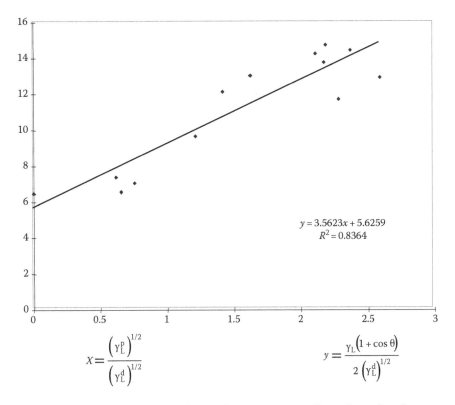

$$X = \frac{\left(\gamma_L^p\right)^{1/2}}{\left(\gamma_L^d\right)^{1/2}} \qquad\qquad y = \frac{\gamma_L\left(1 + \cos\theta\right)}{2\left(\gamma_L^d\right)^{1/2}}$$

Figure 4.11 Surface free energy evaluation for cement paste. (From the authors.)

Figure 4.11 illustrates the evaluation of the surface free energy of cement paste (solid state) by means of a linear regression based on the values of the polar and dispersion components of surface free energies of slurries, distilled water, and reference liquids.

Table 4.4 gives the values of solid surface free energy, in comparison with glass and silicon paper.

It appears clearly that the dispersion interaction is quite larger than the polar one, especially for cement paste with hydrophobic treatment. It is possible to evaluate $\gamma_S^{\text{concrete}}$ by the application of a ratio between the surface of the concrete really occupied by mortar and rock determined by surface topology evaluation. As an example, for a surface constituted with 65% limestone and 35% cement mortar, we obtained (Courard, 2002):

$$\gamma_S^{\text{concrete}} = 0.65\gamma_S^{\text{limestone}} + 0.35\gamma_S^{\text{cement mortar}} = 47.68 \text{ mN/m}$$

Table 4.4 Surface free energy of solids

	Surface free energy (mN/m)		
Support	γ_s^d	γ_s^p	γ_s
Cement paste	31.65	1.69	44.34
Cement paste + hydrophobic treatment	14.86	0.01	14.87
Limestone	37.08	12.40	49.48
Glass	20.54	22.85	43.39
Silicon paper	12.59	5.41	18

Source: Courard, L., *Mater. Struct.*, 35, 149, 2002.

4.5.5 Interpretation and analysis of results

The evaluation of these parameters is fundamental for the study of interface stability; this can be analyzed by calculating the work of adhesion,
interfacial energy, or critical surface energy. Defining the critical surface
energy (γ_C), Zisman proposed a practical approach in order to assess the
wettability of solid surface energies (Figure 4.12). The backbone idea of
this approach is that a solid surface will be wetted by liquids with surface
energies below the critical surface energy of the solid.

Calculation of γ_C for different supports (Table 4.5) gives information
about the opportunity of wetting the mineral surfaces with cement slurries;
no centrifuged solutions of cement modified slurries has a surface energy
lower than critical surface free energy and, consequently, is able to perfectly
wet the concrete surface.

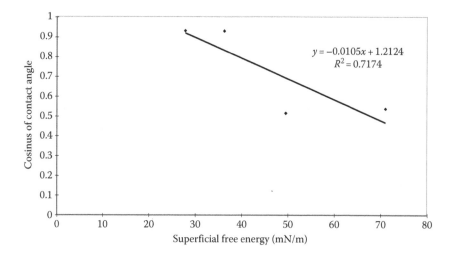

Figure 4.12 Zisman plots for cement paste substrate. (From the authors.)

Table 4.5 Critical surface energy of solids

Support	Critical surface energy, γ_C (mN/m)
Cement paste	25.5
Cement paste + hydrophobic treatment	24.6
Limestone	42.5

But it seems easier for liquids to wet limestone than cement surfaces, which could explain particularly good adhesion between cement mortars and calcareous aggregates. It is also evident that resinous materials (Table 4.6) are developing better adhesion than cementitious surfaces due to a lower surface free energy (Mouton, 2003; Comyn, 1992; Fiebrich, 1994).

Moreover, Dann (1970) demonstrated that "when a liquid series possessing only dispersion forces is used, the critical surface free energy equals the dispersion force component of the surface free energy of the solid and, consequently, provides a method for determining the true surface free energy of the solid." This way of analyzing should be investigated.

Interface stability may be evaluated by the calculation of the work of adhesion that represents the work necessary to separate the support and adherent. If adherence is only due to Van der Waal's forces, the work of adhesion is given by Equation 4.2

$$W_A = \gamma_L + \gamma_S - \gamma_{SL} \tag{4.38}$$

Taking into account the equation of Owens and Wendt (Equation 4.2), evaluation of the work of adhesion is given by Equation 4.39

$$W_A = 2\left(\gamma_L^d \cdot \gamma_S^d\right)^{1/2} + 2\left(\gamma_L^p \cdot \gamma_S^p\right)^{1/2} \tag{4.39}$$

Different values of polar and dispersive components of liquids and solids previously determined led us to calculate W_A (Table 4.7).

Values determined here are based on the interaction between slurries (liquid) and concrete support (solid). It actually does not represent a solid/solid relation, and the work of adhesion does not evaluate the state and intensity of interaction between hardened products. However, it may yield information about the appetency (Courard, 2000) of the liquid for the solid phase and eventual correlation with adhesion properties.

In order to quantify precisely the solidity of the bond, the surface free energies of the bodies to be bonded need to be evaluated, which necessitates sophisticated test devices and measurements (Tabor, 1981). Unfortunately, the available tools are not always well adapted to the particularities of mineral substrates like cement paste and concrete due to problems of roughness, capillary suction, and chemical heterogeneity of the surface.

Table 4.6 Polymers' critical surface energy (mJ/m^2) at 20°C

Polymer	γ_C
Poly(1,1-dihydroperfluorooctyl-methacrylate)	10.6
Polyhexafluoropropylene	16.2
Polytetrafluoroethylene (PTFE)	18.5
Polytrifluoroethylene	22
Polydimethylsiloxane	24
Polyvinylidene difluoride (PVDF)	25
Polyvinyltrimethylsiloxane	25
Polyvinyl fluoride (PVF)	28
Polyvinylmethyl ether	29–30
Polypropylene (PP)	29
Polyethylene (PEBD)	31
Polytrifluorochloroethylene (PCTFE)	31
Polypropylene oxide (PPOX)	32
Polyacrylamide	33
Polystyrene (PS)	33
Polyethylmethacrylate	33
Polyethylacrylate	35
Cellulose acetate (CA)	36
Polyvinyl alcohol (PVAL)	37
Polyvinyl acetate (PVAC)	37
Poly(methyl methacrylate) (PMMA)	39
Polyvinyl chloride (PVC)	39
Polyvinylidene chloride (PVDC)	40
Poly(methyl methacrylate)	41
Polycarbonate (PC)	42
Polyamide-6 (PA-6)	42
Polyurethanes (PUR)	42–45
Polyethylene terephthalate (PET)	43
Epoxy resins (EP)	43–44
Polyamide-11 (PA-11)	43
Polyacrylonitrile (PAN)	44
Poly(1,4-cyclohexanedimethylene-terephtalate)	45
Cellulose	45
Polyamide-6,6 (PA-6,6)	46
Resorcinol–formaldehyde resins	52
Urea–formaldehyde resins (UF)	61

Source: Mouton, Y., *Matériaux organiques pour le génie civil: Approche physico-chimique*, Lavoisier, Hermes, France, 2003, 366p., in French.

Table 4.7 Work of adhesion (mJ/m²)

Liquid	Cement paste	Cement paste + hydrophobic treatment	Limestone	Concrete
Melamine (macromolecules)	99.76	101.46	103.49	102.18
Melamine	102.36	104.08	106.14	104.82
Naphtalene	102.99	104.94	107.15	105.7
Vinyl copolymer	84.04	85.35	86.98	85.95
Maleic acid	100.58	102.28	104.31	103
Natrium lignosulfonate	98.58	100.18	102.13	100.89
Cement slurry (no admixture)	106.69	108.81	111.18	109.61
Water	102.49	104.19	106.23	—

Source: Courard, L., Mater. Struct., 35, 149, 2002.

A large area of investigation is open for a better approach of these parameters and a new way for surface free energy measurement of porous solid surface based on capillary suction itself; the speed of absorption is directly related to the physicochemical interaction between liquid and solid and, consequently, to the contact angle and surface free energy of liquid.

4.5.6 Effect of water

Another consideration is the influence of water as an interfacial phase between the adhering liquid phase liquid and the solid substrate. Water actually modifies the critical values of surface energy and work of adhesion, and the resulting equilibrium depends on the respective values of tension of adhesion.

Coatings are designed to protect concrete against environmental aggression. Water is one of the aggressive parameters that contribute to the deterioration of concrete structures. When water comes between the polymeric system and the concrete substrate (Courard et al., 2003), thermodynamic equilibrium is modified, with respect to particular values of surface free energies of materials. From a theoretical point of view, this means a simple generalization of the law of Young and Dupré (Courard, 2000), relatively to a new liquid–liquid interface (Equations 4.15 and 4.16). A contact angle is in this case a visible effect of the interaction between the two liquids to conquer a solid surface. When equilibrium is attempted and if there is no spreading of one liquid to the detriment of the second one, equilibrium of forces means (Figure 4.13)

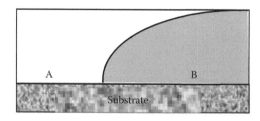

Figure 4.13 Scheme of wetting of a solid surface by two immiscible liquids (wetting is here favorable to liquid B). (From the authors.)

$$\gamma_{SA} = \gamma_{SB} + \gamma_{AB} \cdot \cos\theta \tag{4.40}$$

or

$$\gamma_{SA} - \gamma_{SB} = \gamma_{AB} \cdot \cos\theta \tag{4.41}$$

where γ_{SA}, γ_{SB}, γ_{AB}, and θ are interfacial tensions between the solid S and liquids A and B, interfacial tension between liquids A and B, and the contact angle of these liquids on the solid surface, respectively.

It is possible to show (Kinloch, 1987) that the liquid with the higher tension of adhesion ($= \gamma_x \cdot \cos\theta_x$) will expulse the other one from the surface. The calculation of the work of adhesion (Equation 4.42) can also give interesting conclusions, taking into account the variation of the surface free energies in the presence of water:

$$W_x(L) = \gamma_x(L) \cdot (1 + \cos\theta_x(L)) \tag{4.42}$$

The work of adhesion is an "evaluation" of adhesion, and Table 4.8 clearly indicates the loss of adhesion when water is present at the interfaces between acrylic or epoxy resins and concrete.

Table 4.8 Calculation of the work of adhesion for interfaces without (W_A) and with (W_{AL}) water

Interface	W_A (mJ/m²)	W_{AL} (mJ/m²)	$(W_{AL} - W_A)/W_A$ (%)
Mortar/concrete	87.8	No sense	—
Acrylic/concrete	74.1	22.7	−69.4
Acrylic/acrylic	80.4	53.7	−33.2
Acrylic/hydrophobic treatment	52.2	66.7	27.8
Epoxy/concrete	79.6	21.8	−72.6
Epoxy/epoxy	92.4	53	−42.6
Epoxy/hydrophobic treatment	56	42.2	−24.6

Source: Courard, L., *Mater. Struct.*, 35, 149, 2002.

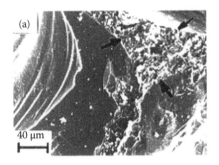

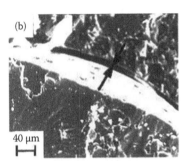

(c)

W_{agg} (%)	$(f_c^w - f_c^{dry})/f_c^{dry}$ (%)	$(f_b^w - f_b^{dry})/f_b^{dry}$ (%)
Polymer concrete		
0.25	−36.8	−22.5
0.50	−48.6	−58.7
1.50	−84.2	−87.5
Polymer mortar		
0.25	−42.2	−24.4
0.50	−43.8	−61.2
1.50	−69.9	−74.5

Figure 4.14 SEM micrographs showing examples of PC microstructure with (a) dry and (b) wet aggregate and (c) the effect of aggregate moisture on compressive f_c and flexural f_b strengths. (From Courard, L. and Garbacz, A., *Restorat. Build. Monum.*, 16(4/5), 291, 2010.)

The water effect is also important in the case of polymer concrete properties. The investigations were carried out for vinyl-ester mortar (A/B = 4 by mass) and vinyl-ester concrete (A/B = 8 by mass) with aggregates of different moisture contents (Wagg—water content by mass). Garbacz and Garboczi (2003) showed that the level of adhesion between resin binder and aggregate affected the mechanical properties of PC (Figure 4.14a and b). As the moisture level of the aggregates increased, the mechanical properties decreased (Figure 4.14c). The variation of these properties with aggregate moisture content is statistically significant—of the same range as decrease in the work of adhesion due to the presence of water. That is why, in the case of polymer composites, a dried aggregate is commonly recommended.

4.6 CONCLUSION

Basic angles of contact measurements are really useful for appraising the quality of the interfacial bond developing between a solid surface (concrete) and a solidifying liquid (repair material, coating, impregnation, etc.). However, for a reliable quantitative evaluation of the adhesive forces generated, the surface free energies of the two materials need to be determined. This is a complex task as many of the available characterization tools are not well adapted for the evaluation of mineral materials like concrete and cement paste, due notably to their heterogeneity, surface roughness, and capillary suction. The presence of water between the adhering material and the solid substrate further increases the complexity of the problem as the liquid with the higher tension of adhesion will expel the other liquid phase from the surface, which for instance is very often the case when water is in concurrence with polymeric materials on the concrete surface.

The fundamental notions reviewed in this chapter regarding interface bond development and thermodynamic equilibrium are prerequisites to the global concept of compatibility presented in Chapter 5.

REFERENCES

Comyn, J. (1992) Contact angles and adhesive bonding. *International Journal of Adhesion and Adhesives*, **12**(3), 145–149.

Comyn, J., Brackley, D.C., and Harding, L.M. (1993) Contact angles of liquids on films from emulsion adhesives and correlation with the durability of adhesive bonds to polystyrene. *International Journal of Adhesion and Adhesives*, **13**(3), 163–171.

Courard, L. (1999) How to analyse thermodynamic properties of solids and liquids in relation with adhesion? In: *Proceedings of the Second International Symposium on Adhesion between Polymers and Concrete (ISAP'99)*, Dresden, Germany (Eds. Y. Ohama and M. Puterman), Rilem Publications, Cachan, France, pp. 9–20.

Courard, L. (2000) Parametric study for the creation of the interface between concrete and repair products. *Materials and Structures*, **33**, 65–72.

Courard, L. (2002) Evaluation of thermodynamic properties of concrete substrates and cement slurries modified with admixtures. *Materials and Structures*, **35**, 149–155.

Courard, L. (2005) Adhesion of repair systems to concrete: Influence of interfacial topography and transport phenomena. *Magazine of Concrete Research*, **57**(5), 273–282.

Courard, L., Darimont, A., Degeimbre R., and Wiertz, J. (2003) Hygro-thermal application conditions and adhesion. In: *Fifth International Colloquium Industrial Floors'03* (Ed. P. Seidler), Technische Akademie Esslingen, Ostfildern/Stuttgart, Esslingen, Germany, pp. 37–142.

Courard, L. and Garbacz, A. (2010) Surfology: What does it mean for polymer concrete composites? *Restoration of Buildings and Monuments*, **16**(4/5), 291–302.

Courard, L., Michel, F., and Martin, M. (2011) The evaluation of the surface free energy of liquids and solids in concrete technology. *Construction Building Materials*, **25**(1), 260–266.

Dann, J.R. (1970) Forces involved in adhesive process. I. Critical surface tensions of polymeric solids as determined by polar liquids. *Journal of Colloid Interface Science*, **32**(2), 302–331.

Derjagin, B.V. (1978) *Adhesion of Solids*. New York: Plenum Publishing Company, 457p.

Fiebrich, M.H. (1994) Scientific aspects of adhesion phenomena in the inter-face mineral substrate-polymers. In: *Proceedings of the Second Bolomey Workshop on Adherence of Young and Old Concrete* (Ed. F.H. Wittman), Aedificatio Verlag, Unterengstringen, Sion, Switzerland, pp. 25–58.

Fox, H.W. and Zisman W.A. (1950) The spreading of liquids on low energy sur-faces. I. polytetrafluoroethylene, *Journal of Colloid Science*, 5(6), 514–531.

Garbacz, A. and Garboczi, E.J. (2003) Ultrasonic evaluation methods applicable to polymer concrete composites. NIST Report No. NISTIR 6975, National Institute of Standards and Technology, Gaithersburg, MD, 73p. http://fire.nist.gov/bfrlpubs/build03/PDF/b03061.pdf.

Gutowski, W. (1985a) Physico-chemical criteria for maximum adhesion. Part I: Theoretical concepts and experimental evidence. *Journal of Adhesion*, 19, 29–49.

Gutowski, W. (1985b) Physico-chemical criteria for maximum adhesion. Part II: A new comprehensive thermodynamic analysis. *Journal of Adhesion*, 19, 51–70.

Gutowski, W. (1987) The relationship between the strength of an adhesive bond and the thermodynamic properties of its components. *International Journal of Adhesion and Adhesives*, 7(4), 189–98.

Harkins, W.D. and Jordan, H.F. (1930) A method for the determination of sur-face and interfacial tension from the maximum pull on a ring. *Journal of American Chemical Society*, 52, 1751–1772.

Kamusewitz, H. and Possart, W. (1985) The static contact angle hysteresis obtained by different experiments for the system PTFE/water. *International Journal of Adhesion and Adhesives*, 5(4), 211–215.

Kinloch, A.J. (1987) *Adhesion and Adhesives: Science and Technology*. London, U.K.: Chapman & Hall.

Michel, F., Perkowicz, S., Courard, L., and Garbacz, A. (2014) Effects of limestone fillers on surface free energy and electrical conductivity of the interstitial solution of cement mixes. *Cement and Concrete Composites*, 45, 111–116.

Mouton, Y. (2003). *Matériaux organiques pour le génie civil: Approche physico-chimique*. Hermes, France: Lavoisier, 366p (in French).

Possart, W. and Kamusewitz, H. (1992) Some thermodynamic considerations con-cerning the usual interpretation of wetting on solids. *International Journal of Adhesion and Adhesives*, 12(1), 49–53.

Possart, W. and Kamusewitz, H. (1993) The thermodynamics and wetting of real surfaces and their relationship to adhesion. *International Journal of Adhesion and Adhesives*, 13(2), 77–84.

Rouxhet, M. (1996) Mécanismes de transfert aux interfaces dans les systèmes multimatériaux. Université Catholique de Louvain. C.R.M.A. Chimie des interfaces, Contrat de recherche Multimatériaux, Région Wallonne (internal reports).

Sasse, H.R. and Fiebrich, M. (1983) Bonding of polymer materials to concrete. *Materials and Structures*, 16(94), 293–301.

Tabor, D. (1981) Principles of adhesion—Bonding in cement and concrete. In: *Adhesion Problems in the Recycling of Concrete* (Ed. P. Kreijger), NATO Scientific Affairs Division, New York, pp. 63–90.

Chapter 5

Compatibility

5.1 GENERAL CONSIDERATIONS

Until recently, a widespread philosophy in the design of concrete repair was based on this simplistic principle: "repair like with like" (Emmons and Vaysburd, 1995). According to this principle, the properties of the repair material should be as close as possible to those of the substrate. The basic idea behind the concept is the following: if the repair material and the old concrete are too different, they may not "work" together and rapid deterioration may occur. Comparative criteria applying independently to individual properties of the repair material and of the substrate were reported by different authors (Emberson and Mays, 1990a; Plum, 1990; Yeoh et al., 1992; Cusson, 1995). Each moderate property mismatch adds to the overall larger mismatch that may lead to cracking and/or debonding. Obviously, cracking of repairs has to be minimized because it threatens not only the durability of the repair, but also the protection of the embedded reinforcing steel.

Though the "repair like with like" approach sounds rational and can be appropriate in some situations, differences in properties or behavior between the base concrete and the repair material will probably always exist, regardless of the material. Even if a deteriorated structure is repaired with a concrete mixture whose proportions and constituents are identical to those of the original concrete, there still will be significant differences since, at the time of casting, the repair concrete is aging and undergoes shrinkage while the base concrete has reached a relative stability in terms of maturity and volume changes. In fact, reported field inspections and tests data indicate that achieving an apparently perfect match of properties does not necessarily prevent the failure of the repair (Emmons and Vaysburd, 1995). Furthermore, as stated by Morgan (1996), "when the existing concrete is of poor quality, the wisdom of repairing it with an equivalent quality concrete must be seriously questioned."

Progressively, a holistic approach based upon the overall *compatibility* between the repair material and the base concrete has gained broad acceptance (Czarnecki et al., 1992, 1999; Emmons and Vaysburd, 1993a,b, 1994; Morgan, 1996; Pigeon and Bissonnette, 1999; Bissonnette and Pigeon, 2000; Courard and Bissonnette, 2008; Vaysburd et al., 2009; Garbacz and

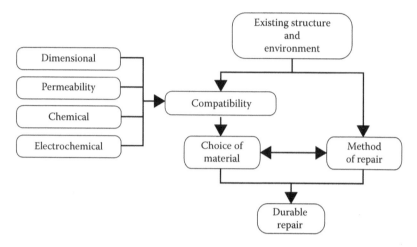

Figure 5.1 Compatibility considerations for a durable concrete repair. (Adapted from Emmons, P.H. and Vaysburd, A.M., Performance criteria for concrete repair materials, Phase I, Technical Report REMR-CS-47, U.S. Army Corps of Engineers, Waterways Experiment Station, Vicksburg, MS, April 1995, 113p.)

Głodkowska, 2011; Garbacz et al., 2014). Following Czarnecki et al. (1992) and Emmons and Vaysburd (1993a) repair compatibility can be defined as:

> the balance of physical, chemical, and electrochemical properties and dimensions between repair material and existing substrate that ensures that the repair withstands all anticipated stresses induced by volume changes, chemical and electrochemical effects without distress and deterioration over a designed period of time.

As summarized in Figure 5.1, four distinct levels of concrete repair compatibility have most generally been considered: *dimensional, permeability, chemical,* and *electrochemical.* In addition to these technical considerations of repair compatibility, a fifth level of compatibility needs to be addressed nowadays in many situations, namely, *aesthetical* compatibility. These different compatibility levels are discussed individually in the following sections.

Although the discussion in this chapter refers more specifically to repairs, it should be made clear that the compatibility concepts apply more globally to all types of interventions on concrete surfaces involving the application of a bonded layer of material, either for protection, repair, or restoration means.

5.2 DIMENSIONAL COMPATIBILITY

Practical experience as described in the technical literature indicates that a durable bond between new and old concrete can be obtained when appropriate care is taken (Felt, 1956; Tabor, 1985; Marosszeky et al., 1991;

Saucier and Pigeon, 1991; Pigeon and Saucier, 1992). However, no repair will be durable unless the repaired layer is free of severe cracking (Marosszeky, 1992) due to the development of internal tensile stresses.

The most important material parameters with respect to the resistance to cracking and/or debonding of a repair are drying shrinkage, creep relaxation, and modulus of elasticity—all dimensional compatibility-related properties (Vaysburd, 2005). Reported failure modes, bond strength data, and long-term performance indicate that repair and overlay of existing structures are a three-phase composite system with an interface transition zone between existing substrate and repair/overlay materials. Long-term bond strength and bond durability depend to a large degree on the quality of the material in the transition zone, on the macro-mechanical interaction in the interface transition zone and on the repair/overlay drying shrinkage.

Harmful tensile and shear stresses arising in the interfacial area of repairs can be generated by external loads, but they are primarily associated with restricted hygrometric and thermal strains. As shown in Figure 5.2 for drying shrinkage, the volume changes in the repair material are partly restrained through the bond by the generally much stiffer support. These movement restrictions, together with the occurrence of a humidity or temperature gradient, induce important tensile stresses that can eventually overcome the tensile strength of the repair material and lead to cracking.

If concrete was a purely elastic material, the achievement of durable crack-free superficial repairs with this material would be somewhat impossible since its ultimate shrinkage strain, which can typically range from 300 to 800×10^{-6} (at 50% R.H.), is much larger than its ultimate tensile strain, which hardly exceeds 150×10^{-6}. In fact, concrete exhibits a viscoelastic behavior, in tension as well as in compression, and shrinkage-induced stresses

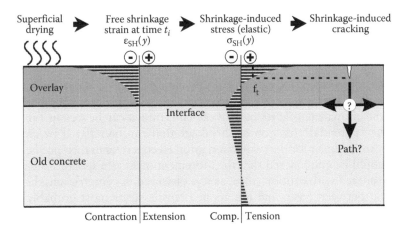

Figure 5.2 Shrinkage-induced stresses and strains in a concrete repair. (From Pigeon, M. and Bissonnette, B., *Concr. Int.*, 21(11), 31, 1999.)

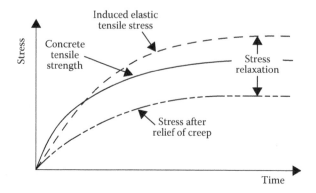

Figure 5.3 Relief of tensile stresses in concrete by creep. (Adapted from Neville, A.M., *Properties of Concrete*, 4th edn., Longham, Essex, England, 1995.)

are relaxed by creep. Cracking will only occur once the actual relieved stress exceeds the tensile strength (Figure 5.3). Thus, it is not the intensity of the individual strain components that matters, but the overall balance of strains. This means that the ultimate free shrinkage value of a given repair material is a useful parameter, in terms of behavior prediction, only as long as other significant properties, such as tensile creep and tensile strength, are known.

While creep in compression has been studied extensively over the years, tensile creep has never been paid much attention to. This can be explained by the fact that concrete testing in tension is quite difficult to achieve properly and also by the fact that properties of concrete in tension are generally disregarded in the design of new concrete structures. Nevertheless, a rather extensive tensile creep testing program was conducted in a research project devoted to the durability of concrete repairs (Bissonnette and Pigeon, 1995). Typical results from this research are presented in Figure 5.4. From this figure, it is obvious that tensile creep can play a significant role in the strain balance mentioned earlier. In the data subset considered here, the plain concrete exhibited less shrinkage than the two fiber reinforced mixtures, but clearly also much less creep. Expressing the raw data of Figure 5.4 in terms of specific creep to shrinkage ratios and extrapolating linearly the creep potential up to the ultimate tensile strength of each mixture, it can be seen in Table 5.1 that the "potential" fraction of shrinkage that can be relieved by creep is significantly higher for the two fiber reinforced concretes in this specific case.

It should be emphasized that the aforementioned extrapolations have to be considered with caution since, as it is observed in compression, the relationship between creep and the applied tensile stress most probably does not remain linear up to failure. Moreover, it is necessary to investigate creep under high stress, just beyond the so-called stress limit above which time-failure will occur, so as to determine the ultimate creep capacity. Nevertheless, in a first approach, the data presented in Table 5.1 indicate

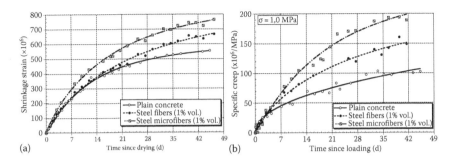

Figure 5.4 Comparative shrinkage and tensile creep shrinkage and tensile creep versus time after 7 days moist curing for plain and steel fiber reinforced concrete mixtures with w (cm) = 0.55: (a) drying shrinkage of plain and fiber reinforced concrete mixtures tested at 7 days (50% R.H.) and (b) total tensile creep of plain and fiber reinforced concrete mixtures tested at 7 days (50% R.H.). (From Bissonnette, B. and Pigeon, M., *Cement Concr. Res.*, 25(5), 1075, 1995.)

Table 5.1 Extrapolation of the potential fraction of restrained shrinkage-induced stress that can be relieved by creep using the specific tensile creep to shrinkage ratio

Concrete mixture[a]	(A) $\dfrac{\text{Specific tensile creep}}{\text{Shrinkage}}$ (MPa^{-1})	(B) Tensile strength (MPa)	(C = A × B) Potential fraction of stress relieved by creep (%)
Plain concrete	0.17	4.4	75
Steel fibre concrete	0.24	4.8	115
Steel microfibre concrete	0.26	3.7	96

Source: Bissonnette, B. and Pigeon, M., *Cement Concr. Res.*, 25(5), 1075, 1995.

[a] For all three mixtures, ASTM type I cement was used and W/C = 0.55; for both fiber reinforced concrete mixtures, the fiber content was 1% by volume.

that the cracking tendency of a material under drying conditions cannot be evaluated solely on the basis of free shrinkage tests.

In addition to shrinkage and creep, the two other properties to consider for dimensional compatibility of repair materials are the modulus of elasticity and the thermal expansion coefficient. It has been demonstrated (Emberson and Mays, 1990b,c; Marosszeky, 1992) that sharp differences in elastic moduli and coefficients of thermal expansion between the repair material and the substrate can be detrimental, depending on factors such as the loading conditions during and after the repair, the creep capacity of the repair material, the bond strength, and the evolution of the temperature distribution within the element over time.

The dimensional compatibility properties and considerations are summarized in Figure 5.5.

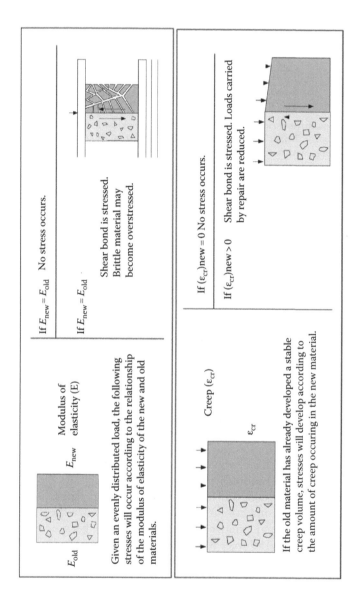

(Continued)

Figure 5.5 Material properties and phenomena influencing the dimensional compatibility of concrete repairs.

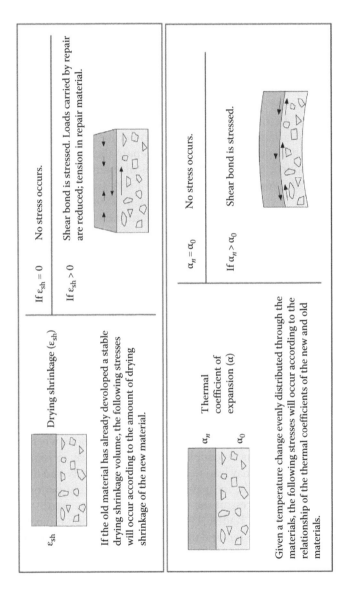

Figure 5.5 (Continued) Material properties and phenomena influencing the dimensional compatibility of concrete repairs. (From ICRI Guideline No. 03733, Guide for selecting and specifying materials for repair of concrete surfaces, International Concrete Repair Institute (ICRI), Des Plaines, IL, January 1996, p. 34.)

5.3 PERMEABILITY COMPATIBILITY

Superficial repairs of concrete structures are often intended to restore adequate protection against the penetration of aggressive agents. Permeability is thus another very important consideration in the choice of a repair material (Vaysburd, 2006). Depending on the type of degradation or the risk of degradation being considered, different philosophies have evolved over the years in the repair industry, advocating either the use of low permeability materials or materials with a permeability as close as possible to that of the original concrete.

If the use of low-permeability concrete in new construction is the key to achieving durability, this rule does not necessarily apply to concrete repairs and overlays, where the situation is significantly more complex. In fact, it is actually wrong in a number of cases and, sometimes, it may end up in a situation worse than prevailed before the repair (Schrader, 1992a,b,c; Emmons and Vaysburd, 1993a, 1995). A typical example of that is the partial resurfacing of a concrete structure with a very tight repair material, such as latex-modified concrete, to repair damage caused by deicing salt penetration. If the penetration of water and salt in the substrate is not fully prevented, chlorides will continue to migrate toward reinforcement and accumulate beneath the impermeable layer. It may result in a more severe corrosion and accelerate the degradation of the structure.

In the same manner, repairing a damaged structure in contact with water using a low-permeability material is likely to cause saturation in the substrate and make it more vulnerable against frost action, especially if the old concrete is of poor quality and has a deficient air void network. Furthermore, the bond between materials having different permeabilities acts as a physical barrier where harmful pressures can be generated during freezing (Schrader, 1992a,b,c).

Emmons and Vaysburd (1993b) asserted that the use of low-permeability repair materials regardless of repair specifics can lead to unsuitable choices, incompatibility problems, and eventual repair failures. According to them, durability of the repair can be negatively affected in many situations where repair and substrate have different, incompatible permeability. Hence, in some cases, the low permeability concept may lead to a false sense of security and the specification of unsuitable materials, incapable of providing lasting performance. A practical example of unsuccessful use of a low-permeability repair material was reported by Schrader and Kaden (1994). In a rehabilitation project, latex-modified shotcrete was used to repair damage triggered by deicing salts around a pier cap. Unfortunately, the top of the cap was not protected and the source of the salt and moisture penetration was not eliminated. In these conditions, an even more severe attack of the reinforcement with subsequent steel corrosion and spalling could develop. Water contaminated with deicing salts dripped from the bridge deck onto the pier cap, penetrating into it with no possibility to escape. Without any intervention, continued deterioration could have been expected, but with such a repair, it was actually accelerated and intensified.

The lesson from the previous example is that in different situations, the selection of low-permeability repair materials that are not compatible with existing concrete may lead to failure. It is important to note that cracking in the repair and/or its debonding can drastically offset the benefit of having a very low-permeability repair material. Microcracks connected with wider cracks originating from the repair surface may play a much greater role in reducing the permeability and durability than the permeability of repair material itself.

Encapsulation of concrete is a special problem with regards to permeability compatibility (Emmons and Vaysburd, 1993b). As a specific example of that problem, numerous bridge columns have been repaired and/or protected in the 1980s and 1990s in North America with vapor-barrier type systems that result in encapsulation of concrete. As the temperature drops, moisture in vapor form migrates toward the barrier and converts into liquid form at the dew point. Water solubles in the concrete are carried along in this migration process. Liquid will then convert into ice at freezing temperatures, resulting in freeze–thaw damage at the edge of the vapor barrier. When this action is reversed by an upswing in temperatures, moisture reconverts back to vapor, leaving water solubles behind in a crystalline form. Repeated cycles can eventually lead to severe deterioration from either one or a combination of these damaging forces.

Although a tremendous amount of information can be found in the literature on corrosion of steel in concrete, comparatively little work has been specifically devoted to the mechanisms of corrosion involved in repaired concrete structures and their most influential factors. Theoretical appraisal of this issue has traditionally focused on differences of electrical potential between the repaired areas and the surrounding nonrepaired substrate. In contrast with the durable repair philosophy discussed in the previous paragraphs, this has led to repair approaches promoting the use of repair materials having permeability characteristics close to those of the substrate concrete in order to keep the difference of potential between the adjacent media as low as possible.

Hanley et al. (1998) have studied the influence of permeability on corrosion activity in concrete repair, addressing both the potential theory and the kinetic theory. Based on different electrochemical measurements, it was clearly shown that this is the change in the anodic reaction in the patch that causes the difference in potential. The permeability of the patching material does not affect the anodic reaction, but rather by controlling the diffusion of oxygen and transport of hydroxyl ions within the patch, it affects the cathodic reaction. By selecting a repair material with similar permeability to the substrate, it can be assumed that the cathodic reactions in both the patch and the substrate will be similar. Hence, the practice of selecting repair materials with equal or higher permeability than the base concrete substrate in an effort to reduce the potential difference between the patch and substrate seems to be a faulty approach (...), as the cathodic reactions in both the patch and the substrate will be similar. The findings of Hanley et al. (1998) suggest indeed that repair materials that are less permeable should have less impact on the surrounding substrate, as well as provide a higher level of protection in the patched area.

Obviously, more work is needed on defining what degree of permeability of repair materials shall be recommended for different repair situations. Most likely, there is no universal recommendation as to whether very low, or compatible with existing concrete permeability materials, are more effective. It depends for instance on particular transport mechanisms in the repair system. Transport of substance through and in the repair system is a very complex process, consisting of a combination of liquid flow through macro- and microcrack systems, capillary transport, diffusion, and osmotic effects. The exact contribution of each process can vary depending on the particular repair situation. The effects of variables such as location of the repair in the structure, internal environment in the repair system, amount and distribution of cracks in both phases of the composite repair system, temperature, moisture, and stresses need to be considered.

Therefore, while it is clear that a low permeability is highly desirable as far as durability of concrete as an individual component is concerned, using a low-permeability repair material regardless of the conditions is a mistake. In case of repairs there is no *"rule of thumb"*; each situation is different and requires to be accurately analyzed. The most significant aspects to take into account include permeability of the substrate, potential access of moisture, salts and other aggressive agents, and the exposition to freeze and thaw cycles.

5.4 CHEMICAL COMPATIBILITY

Chemical compatibility implies that the repair material will not have any adverse effects on the substrate it is meant to bond to and protect.

Repair failures due to the chemical incompatibility between a repair material and concrete substrate are very seldom reported in the literature and are being practically ignored by engineers and scientists. Nevertheless, chemical incompatibility is taking place more often than acknowledged, and it is often a contributing factor to other major causes of repair failures. Emmons and Vaysburd (1993b) addressed the issue, stating that repair materials specified and used for repair jobs should be chemically compatible with the existing concrete substrate to avoid premature repair failures. Chemical compatibility properties to consider, according to authors, may include alkali content, C_3A content, chloride content, sulfates content, etc. All aspects of chemical compatibility need to be considered in the selection of repair materials. For instance, when concrete being repaired is affected by alkali-aggregate reactivity (AAR) or incorporates potentially reactive aggregates, the repair material alkali content should be strictly controlled.

The reactivity of the repair material to reinforcing steel and other embedded metals or to specific protective coatings or sealers applied over the surface repair must also be considered. Repair materials with moderate to low pH may provide little protection to reinforcement. Moreover, certain repair materials are not compatible with waterproofing membranes required as protection, following a repair. Therefore, the reactivity of the various repair materials with both the substrate components and/or surface protection product should be considered.

When designing a repair, the parameters to be addressed may differ depending upon whether carbonation or chloride contamination is the main cause of corrosion and deterioration. For instance, in the rehabilitation of a chloride-contaminated reinforced concrete structure involving undercutting of reinforcement and reembedment with a repair material, the potential movement of chlorides between the existing concrete and the new material must be considered. Repaired areas, which are chloride-free at the time of repair, may well reach a high chloride content after some time. As an example, we may consider an overlaid slab-on-grade exposed to dry, windy weather, high temperature and very low relative humidity, with moisture condensing from the bottom of the slab. Progressively, water and ions such as alkalis and chlorides start to migrate to the top of the slab, resulting in the overlay contamination.

5.5 ELECTROCHEMICAL COMPATIBILITY

Corrosion of steel reinforcement is one of the most important durability problems of concrete. In a rehabilitation project, whenever the structure to be repaired is reinforced, the electrochemical compatibility must be considered.

The driving force for the phenomenon of corrosion in repair systems is in fact generally attributed to the electrochemical incompatibility between the repair and the existing concrete. Electrochemical incompatibility can be defined as the imbalance in electrochemical potential between the reinforcing steel located in the repair area and that located in the nonrepaired areas, as a result of dissimilar local environments (Gu et al., 1997). The dissimilarities can relate to the physical properties, chemistry, and/or to the internal environment of the respective embedding materials. In that regard, Emmons and Vaysburd have listed the main differences between new construction and repair as a guide for proper selection of repair strategies (Emmons and Vaysburd, 1997).

Still, the issue is rather complex and controversial. In some investigations (Gulikers and Van Mier, 1991; Marosszeky and Wang, 1991), the reported data tend to indicate that polymer-based and polymer-modified repair materials, which have a dense microstructure and a high electrical resistivity, provide a more reliable protection against corrosion than ordinary Portland cement repair materials. On the contrary, it is suggested in some other reports (Emmons and Vaysburd, 1993a, 1994; Dehawah et al., 1994) that repair materials with composition and permeability similar to those of the surrounding concrete should be used to ensure or at least optimize electrochemical compatibility. According to Emmons and Vaysburd (1993a, 1994), highly resistive materials tend to isolate the repaired area from the adjacent undamaged concrete and if there exists a significant permeability or chloride content differential, corrosion is confined in a restricted area, its rate may increase and cause premature failure in either the repair or the contiguous concrete.

Vaysburd et al. (2009) addressed the relationship between the durability of repaired structures and the electrochemical compatibility issue. To provide adequate resistance to aggressive actions, it is necessary to foresee how the repaired structure will deteriorate. In other words, the aggressiveness of the existing internal (i.e., inside the structure) and exterior environments, their interaction, and the possible changes caused by the repair should be given comprehensive consideration at the design stage. Such a thorough analysis is necessary to achieve electrochemical compatibility and fulfill the requirements for durability and structural safety of aging infrastructure. Nevertheless, the authors recognize that among the durability/service life problems of concrete repair, the issues relating to electrochemical compatibility are the least understood and that a precise appraisal is very difficult, if not impossible.

The difficulties are mainly due to the following three factors:

- The existing structure has its unique internal environment, as influenced by aging, weathering, and chemical/electrochemical changes and activities, which necessitated the repair.
- While the exterior environment of a structure depends essentially on its geographical location (e.g., temperature, humidity, wind, radiation, soil characteristics, etc.) and the human activity nearby (industrial- or traffic-generated pollution), the local internal environment is altered by the presence of a repair.
- In repair systems, the internal environment is like a moving target, constantly changing due to the existence of internal transport mechanisms. Actually, the altered local environment promotes the movements of water and ions driven by temperature gradients, pressure gradients, moisture gradients, concentration gradients, and/or the presence of an electric field.

A major problem with many repaired concrete structures is the re-inception, and even sometimes acceleration, of reinforcing steel corrosion shortly after rehabilitation. When chloride-contaminated reinforced concrete is repaired, some of the contaminated concrete will often be left in place. As a result, the repair material has moisture, oxygen, and chloride contents different from that of the surrounding concrete. Strong corrosion cells may develop, resulting in spalling of the repair itself or of the surrounding concrete, the latter being typically referred to as the *halo* or *ring* effect.

The combined effects of the several critical dissimilarities in environment along the electrically continuous steel rebar (in addition to a variable stress state) significantly add to the overall complexity of the problem. When steel in the repair area is only partially exposed, with a portion of its length left embedded in chloride-contaminated concrete and the other portion re-encapsulated in a repair material, strong corrosion cells may develop. The portion of the bar located in the existing concrete becomes anodic, corroding at a rapid rate, driven by the other portion embedded in the repair material, which acts as the cathode. Repair deterioration and failure may develop within months.

If the existing concrete is completely removed from around the reinforcement and replaced by a repair material, similar reactions can accelerate steel corrosion at the perimeter of the repair, in the surrounding areas of existing concrete.

When encasing the reinforcing steel of a chloride-contaminated structure in a repair material, the possible movement of chlorides within the repaired structure must be considered. In the example referred to previously, where an overlaid contaminated slab was subject to moisture condensation from the bottom of the slab owing to the exposure conditions, the repair layer was contaminated after a while due to upward movement of water and its ionic content, notably the chlorides. Unless the repair project provides for global cathodic protection, the risk of re-incipient corrosion in repaired structures still containing chloride-contaminated areas is significant.

Among various factors affecting compatibility in concrete repair systems, electrochemical compatibility is the most complex and critical for adequate performance. It is difficult to predict the effect of repair upon the overall electrochemical activity in a rehabilitated concrete structure, because it is a function of wide variety of parameters. The risk of recurring corrosion, and even its acceleration, due to electrochemical incompatibility between the old and new portions of the repaired structure is most generally significant, unless cathodic protection is implemented. Unfortunately, the current state of knowledge does not allow addressing adequately the issue of electrochemical compatibility on a systematical basis, hence affecting the ability to predict the future service life of a repair concrete structure.

5.6 AESTHETICAL COMPATIBILITY

Repairs of historic buildings, monuments, and architectural concrete usually have to meet aesthetical compatibility requirements, especially with respect to color, finish, and texture of the existing concrete to be restored. Typically, this cannot be achieved with the use of materials such as surface coatings to hide the repaired areas and obtain a uniform aspect, as the original concrete must generally be left in its original condition and aspect. Aesthetical compatibility of repair can be accomplished by using specialty materials such as colored concrete, polished concrete, exposed-aggregate concrete, etc., as well as adapted surface preparation and placement techniques.

In many practical situations, the most challenging aspect of aesthetical compatibility is to match as close as possible the original color of the structure or monument, allowing for the variations from one area to the other. A recently developed technology referred to as the *liquid pigment automated dispensing* concrete (Forgey, 2005) is rising as a quite promising solution for aesthetical compatibility of repairs in terms of color and tint. Much like a computer printer, a typical liquid-dispensing system installation is equipped with different tanks of liquid primary colors, from which wide palettes of colors and tints can be prepared on demand. The level of precision offered by this innovative technique provides users with the flexibility and accuracy required to overcome the constraints encountered in the field and satisfy the often very stringent demands of architects, artists, curators, heritage managers, and owners in general.

5.7 ACHIEVEMENT OF COMPATIBILITY IN REPAIR SYSTEMS

Compatibility and durability design of repair systems does not necessarily involve calculations comparable with those carried out for structural design and safety. Rather, much more emphasis falls upon both conceptual analysis and adapted specifications. Primarily intended to improve the resistance to deleterious environmental factors and the risk of premature failure of the repaired structure, the compatibility approach can also simplify and improve the field operstions.

Significant advances are still to be made in adequately matching different types of repair materials and OPC concrete to ensure acceptable long-term service. A better understanding of the fundamental properties and characteristics will help prevent premature failures, lead to greater composite durability, and, therefore, pave the way for innovation in materials and applications.

A clearer appreciation is needed of those physical properties that may provide the key to successful and durable use of a given material in a given situation. An integral part of this appreciation is the assessment of the likely consequences of the property mismatches (e.g., CTE, modulus of elasticity, and creep). Underlying this assessment is the further need to make a clear judgment of the relevance of the material property test data and the methods used to obtain them.

There are always property differences, mismatches between any repair materials, and concrete substrates, which may lead to application and performance problems. Understanding and recognizing these differences are crucial to the success of a repair project. The choice of the most suitable material for a given repair application is, of necessity, a compromise between material properties, exposure conditions, substrate conditions, budget constraints, etc. Selection should always be based on as much knowledge of the relevant properties as possible.

It is important to stress that our concern with compatibility should neither be based solely on the materials themselves, nor on the applications they are used for. In fact, the concept of compatibility is all about achieving the optimal middle ground solution to obtain a lasting monolithical action of the composite repair system.

5.7.1 Experimental characterization of compatibility parameters

5.7.1.1 Dimensional compatibility

The issue of dimensional compatibility relates to the mechanical balance within the composite system made of existing concrete and a bonded surface treatment, as influenced by differences in terms of mechanical properties and volume changes.

The respective mechanical stiffness characteristics (elastic modulus, Poisson's ratio, and creep) of the old concrete and the surface treatment determine their load-sharing ability and the risk for stress concentrations in the composite.

The problems associated with differential volume changes in concrete repairs are triggered by the restraint provided by the existing concrete and the reinforcement. It results in a complex interplay, or competition, between the tensile stresses induced by the restrained strains, the stress relief due to creep, and the gain in strength, which determines whether or not cracking and/or debonding will occur.

Test methods for the evaluation of repair material properties and characteristics related to dimensional compatibility are listed in Table 5.2.

Evaluating the cracking sensitivity of a repair material based upon free shrinkage data alone is very questionable, since the volume changes in repairs are in fact restrained by the substrate and sometimes supplemental reinforcement. Restrained shrinkage tests have been developed to evaluate the restrained volume change behavior of cement-based materials, notably repair materials, and their sensitivity to cracking. The following types of restrained shrinkage tests have been traditionally used for cement-based materials (Emmons et al., 2000):

- Linear test, where the restraint is typically provided either internally by embedded metallic bars or externally by a rigid metallic frame;

Table 5.2 Common types of tests for assessing the dimensional compatibility properties of repair materials

Description	Standard test method(s)
Modulus of elasticity and Poisson's ratio	ASTM C469; ASTM C580
	ISO 6784; NBN B15 002
Creep	ASTM C512; ASTM C1181
Length change/drying shrinkage (free)	ASTM C157; ASTM C596
Restrained shrinkage	ASTM C1581; AASHTO PP34
Restrained expansion	ASTM C806
Thermal expansion	ASTM C531; ASTM C884
	ASTM D696; CRD-C39

- Plate test, where the restraint is typically provided by a metallic plate onto which the material is placed;
- Ring test, where the restraint is provided by an inner metallic ring around which the material is cast.

Vaysburd et al. (2000) developed a test procedure using test slabs with a cavity (sometimes referred to as *"box test"*) to assess the cracking sensitivity of repair materials in representative conditions (Figure 5.6a) During and after conditioning, the repaired test slabs are monitored closely for cracking (number, width, length, location), which allows to assess dimensional compatibility on a comparative basis.

Czarnecki et al. (1992) used a cavity test specimen developed for other means (Figure 5.6b) by RILEM TC 124-SRC (1994) to assess the compatibility of repair systems (Figure 5.6b). After proper pre-conditioning, the concrete beams are filled with a repair material to be tested, as in the procedure proposed by Vaysburd et al. (2000), and then submitted to curing and conditioning. Ultimately, the repaired test specimens are tested in flexure (4-pt. bending). The failure mode observed during the flexural experiment is used for the evaluation of the repair system's compatibility, which can be classified as either adequate or inadequate (Table 5.3).

5.7.1.2 Permeability compatibility

While low permeability is often looked after in the design and selection of repair materials and treatments for concrete, engineering analysis and judgment are needed to define what degree of permeability should actually be recommended for different repair situations. Most likely, there is no systematic answer on whether very low permeability materials, or compatible permeability materials, should be opted for.

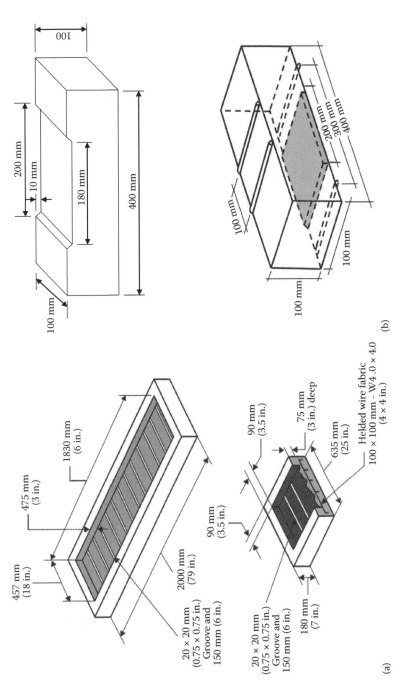

Figure 5.6 Test specimens used for the assessment of compatibility of repair materials: (a) Box test proposed by Vaysburd et al. (2000) and (b) specimen with cavity used for corrosion evaluation, as reported by RILEM TC 124-SRC (1994).

Table 5.3 Typical classification for PC/PCC mortar to concrete bond depending on the type of failure observed in composite test specimens subjected to 4-pt. bending acc. to RILEM TC 124-SRC

Compatibility assessment	Failure mode	Description	Examples
Adequate	A: B:	A and B: Adequate mortar-to-concrete adherence over the whole interface with concrete	K-3/4B V-8/4B
Inadequate	C:	Inadequate mortar-to-concrete adherence on one of the inclined interfacial planes and partial debonding of the connecting horizontal interfacial plane	T-2/1B
	D:	Inadequate mortar-to-concrete adherence on all interfacial planes	M-5/D

Source: Czarnecki, L. and Runkiewicz, M., On the compatibility measure in the repair systems, in: H. Beushausen, F. Dehn, and M.G. Alexander, eds., *Proceedings of the International Conference on Concrete Repair, Rehabilitation and Retrofitting (ICRRR'2005)*, Balkema, Rotterdam, the Netherlands, 2005, pp. 1003–1008.

Table 5.4 Common types of tests intended to characterize the properties or mechanisms governing the movement of moisture in cementitious materials

Description	Standard test method(s)
Water permeability of concrete	CRD-C 48-92
	CRD-C 163-92
	Figg's method
Air permeability of concrete	*Figg's* method
Determination of the volume of permeable voids	ASTM C39
Migration/diffusion tests	ASTM C1202
	NT Build 443; NT Build 492
Capillary absorption	ASTM C1585; ASTM C1757
	EN 1062-3

Transport of substances through and in the repair systems is a complex process, consisting of a combination of capillary transport, diffusion, osmotic effects, and liquid flow through cracks. The actual contribution of each mechanism vary from one situation to another, depending primarily on the respective porosity and cracking characteristics (volume, size distribution, and connectivity) in the existing concrete and repair material, but also on a range of variables such as temperature, moisture, and the chemical environment within the composite repair system.

Different types of test methods intended to characterize the various mechanisms and properties governing the movement of moisture in cementitious materials are listed in Table 5.4. Their relevance needs to be evaluated, based on the materials and conditions.

Over the last couple of decades, permeability criteria based on ASTM C1202 (*Standard Test Method for Electrical Indication of Concrete's Ability to Resist Chloride Ion Penetration*) have been increasingly used in the specifications of many authorities [72]. Nevertheless, the representativeness of this type of test is often being questioned, as the results may be influenced by many parameters other than those related to moisture transport itself.

5.7.1.3 Chemical compatibility

The main chemical compatibility properties or characteristics to be considered in the selection of repair materials are the alkali content, the chloride content, the sulfate content, and the pH. Related standard test procedures are listed in Table 5.5.

5.7.1.4 Electrochemical compatibility

When repairing an existing concrete structure that suffers from corrosion of reinforcement and related concrete deterioration, the risk of continuing

Table 5.5 Tests to evaluate the chemical compatibility of repair materials (ACI 364.3)

Description	Standard test method(s)
Total sulfur trioxide (SO_3) content	ASTM C114
Total alkali content	ASTM C114
Chloride content	ASTM C1152
	ASTM C1218
pH	ACI 364.3R-09

corrosion due to the electrochemical incompatibility between "old" and "new" is always present, unless global cathodic protection is implemented. In evaluating that risk, the influence of the repair phase on the existing phase, change in chemical composition, distribution of aggressive agents, oxygen, moisture, and other factors on the electrochemical properties of the repair system should all be considered.

Some of the most common types of tests for assessing steel corrosion risks, activity, and extent in an existing reinforced concrete substrate as well as characterizing the ability of the repair material to resist subsequent contamination are summarized in Table 5.6, together with related standards. The information generated with the various test methods listed can be used in evaluating the requirements for achieving electrochemical compatibility of repair. Nevertheless, no guidance for doing so is available yet. Corrosion of steel in concrete is by itself an intricate problem involving a variety of phenomena and establishing electrochemical compatibility is a challenge that essentially relies on experience and empirical knowledge.

Table 5.6 Common types of tests for assessing steel corrosion risks, activity, and extent in an existing reinforced concrete substrate, as well as characterizing the ability of the repair material to resist subsequent contamination

Repair component to be assessed		Description	Standard test method(s)
Existing concrete	Repair material		
X		Measurement of corrosion potential	ASTM C876
X		Determination of chloride content distribution	ASTM C1152
X		Measurement of carbonation depth	CPC-18
X		Determination of moisture condition of the concrete	ASTM F2170
X	X	Determination of the volume of permeable voids	ASTM C39
X	X	Migration/diffusion tests	ASTM C1202
			NT Build 443
			NT Build 492

5.7.2 Modeling tools for repair compatibility

Repair materials compatibility need to be considered in light of the various requirements commanded by the actual service conditions. These requirements may be addressed in a mathematical model taking into account the various types of loading to which the repair may be submitted (including gravity loads, temperature and humidity changes, and other chemically- or physically-induced loads) and intended to evaluate a so-called *compatibility space*, as proposed by Czarnecki et al. (1992). In the proposed approach, the model is defined by a set of linear and nonlinear equations (inequalities), where the variables are the relevant properties and characteristics of the repair material and concrete substrate, respectively. The assessment of the compatibility space relies on the resolution of N equations (requirements for compatibility) by successively ascribing eligible values $(x_1, x_2, \dots, x_n)$ to these properties and characteristics. The solution for the set of equations defines the N-dimensional compatibility space.

Czarnecki et al. (1992) and Głodkowska (1994) have proposed different compatibility models for injection, patch repair, and protective coating, respectively. Each model consists of a number of inequalities defining the minimum compatibility requirements for selected repair systems. An example of such requirements is provided in Table 5.7 for polymer composite mortars. The range of values for the material parameters needs to be defined for the given repair type and the level of compatibility can be quantified over the whole domain of variation of these parameters (Table 5.8). Hence, the compatibility space can be advantageously exploited in the design and development of new materials by defining the actual value ranges to be achieved for the different properties.

The mathematical expressions for each parameter are formulated using simplifying assumptions, such that the compatibility model is simple to use, yet generating reliable results. In the case of polymer repair materials for instance, the following simplifying assumptions are made: purely elastic behavior for the repair material; no coupling effects between the various types of loading; and service conditions such that the temperature is kept below the glass transition temperature of the polymers being used. The proper repair material has to be selected, based on the existing concrete susbtrate characteristics and the projected service conditions.

To determine the compatibility space, computer programs were first developed at the Koszalin University of Technology (Głodkowska, 1994, 2003) and the Warsaw University of Technology (Czarnecki et al., 1999). The latest developments from these groundbreaking initiatives led to the development of ANCOMP, a state-of-the-art modeling tool for the evaluation of repair compatibility (Garbacz, 2013). An example of compatibility subspace determined with ANCOMP is presented in Figure 5.7.

Table 5.7 Dimensional compatibility model developed for polymer composite—portland cement concrete substrate (PC-CC) systems

$\varepsilon_{tp} \cdot s_r \geq \Delta w$	(1)	+a	+b	+
$\dfrac{f_{tp} \cdot h_p}{f_{tc}} \cdot \varepsilon_{tp}(1+\varepsilon_{tp}) \geq \Delta w$	(2)	−	+b	+b
$\dfrac{f_{tp} \cdot h_p}{f_{tc}} \cdot \varepsilon_{tp}(1+\varepsilon_{tp}) \geq w_{max} + \Delta w$	(3)	−	+c	+c
$f_{Ao}^{p/b} \geq f_{tc}$	(4)	−	+	+
$f_{As}^{p/b} \geq f_{tc}$	(5)	+	+	−
$f_{Ao}^{pi/pi+1} \geq f_{tc}$	(6)	−	+	+
$f_{As}^{p/b3} > \dfrac{(\alpha_{Tp} - \alpha_{Tc})}{E_{tp} + E_{tc}} E_{tp} \cdot E_{tc} \cdot \Delta T^c$	(7)	+	+	+
$f_{As}^{p/b3} \geq \dfrac{(\varepsilon_{tp} - (f_{tc}/E_{tc}))}{E_{tp} + E_{tc}} E_{tp} \cdot E_{tc}$	(8)	+d	+	+
$\dfrac{\lambda_c}{\lambda_p} < \dfrac{E_{tp} \cdot \alpha_{Tc}}{B}$ $B = \dfrac{(\alpha_{Tp} - \alpha_{Tc})}{E_{tp} + E_{tc}} E_{tp} \cdot E_{tc}$	(9)	−	−	+
$f_{tp} \geq \dfrac{0.3 \cdot E_p \cdot \varepsilon_s}{(1-\nu_p)}$	(10)	+	+	+
$f_{Ao}^{p/b3} \geq \dfrac{0.3 \cdot E_p \cdot \varepsilon_s}{(1-\nu_p)}$	(11)	+	+	+
$f_{Ao}^{pi/pi+1} \geq \dfrac{0.3 \cdot E_p \cdot \varepsilon_s}{(1-\nu_p)}$	(12)	−	+	+
$h_p \geq \pi\sqrt{D \cdot t}$	(13)	−	−	+

Sources: Czarnecki, L. et al., Problem of compatibility of polymer mortars and cement concrete system, *International Colloquium Materials and Restoration*, Esslingen, Germany, 1992, pp. 964–971.

+, Condition that determines the compatibility of a polymer composite with Portland cement concrete and −, the condition does not occur.

a In this case: $l_t = a_r$.
b With regard to the cracked concrete surface.
c With regard to the noncracked concrete surface-cracks appear during service.
d In case of crack injection $f_{tp}^{p/b}$ is used.

Table 5.8 Example of the material properties and characteristics taken into account in a model for the evaluation of dimensional compatibility of repairs

Material property	Repair material	Concrete substrate
Tensile strength (MPa)	f_{tp}	f_{tc}
Modulus of elasticity (in tension) (MPa)	E_{tp}	E_{tc}
Modulus of elasticity (in compression) (MPa)	E_{cp}	E_{cc}
Coeff. of thermal expansion (1/K)	α_{Tp}	α_{Tc}
Coeff. of thermal conductivity (W/mK)	γ_{Tp}	λ_{Tc}
Elongation at break (mm/mm)	ε_{tp}	—
Poisson coefficient (—)	ν_p	—
Max. crack width upon failure (mm)	w_{max}	—
Crack width (mm)	—	w_d
Crack width change (mm)	—	Δw
Drying shrinkage (mm/mm)	ε_s	—
Repair thickness (mm)	h_p	—
Interlayer adhesion (MPa)	$f_{Ao}^{pi/(pi+1)}$	—
Adhesion to the substrate in shear (MPa)	f_{As}	
Adhesion to the substrate in tensile (MPa)	f_{Ao}	
Temperature gradient during service (K)	$\Delta T°$	

Sources: Czarnecki, L. et al., Problem of compatibility of polymer mortars and cement concrete system, *International Colloquium Materials and Restoration*, Esslingen, Germany, 1992, pp. 964–971.

In the example shown, the calculations were performed for a repair system made of a proprietary polymer mortar placed on a 15 MPa concrete substrate. The temperature gradient during service was assumed to reach 30°C. The point corresponding to the combination of mortar properties is located outside of compatibility space. If for instance the elastic modulus of the material could be reduced with the other parameters being kept constant (with proper mixture design modifications), it would result in a larger compatibility space and the mortar would fulfill requirements of compatibility. For the same repair system, the influence of crack width for cracked and noncracked concrete substrate is shown in Figure 5.8.

The concept of compatibility was successfully validated in further research projects (e.g., Czarnecki et al., 2004; Garbacz and Głodkowska, 2012). Based on an extensive experimental program, Głodkowska has proposed modifications to the compatibility requirements, taking into account long-term loads (Głodkowska, 2011) as well as the effect of repair weathering (Głodkowska and Staszewski, 2007).

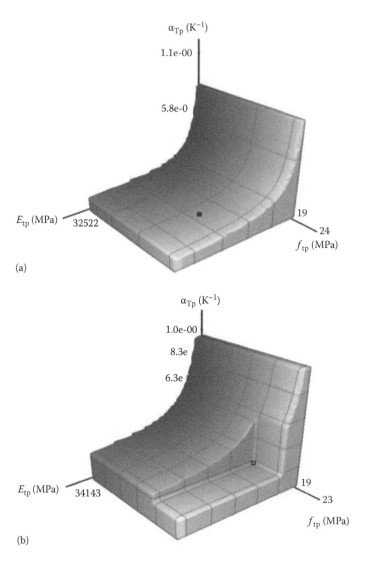

α_{Tp} (K^{-1})

1.1e-00

5.8e-0

E_{tp} (MPa)   32522

19

24

f_{tp} (MPa)

(a)

α_{Tp} (K^{-1})

1.0e-00

8.3e

6.3e

E_{tp} (MPa)   34143

19

23

f_{tp} (MPa)

(b)

Figure 5.7   Examples of compatibility subspaces determined with the ANCOMP code taking into account the elastic modulus E_{tp}, the tensile strength f_{tp}, and the coefficient of thermal expansion α_{Tp}: (a) proprietary polymer mortar with high elastic modulus (incompatibility) and (b) proprietary polymer mortar with similar properties, but a lower elastic modulus (compatibility). (Adapted from Garbacz, A. et al., Patch repair: Compatibility issues, *Proceedings of the Conference Concrete Solutions*, September 1–3, Belfast, U.K., 2014, pp. 231–236.)

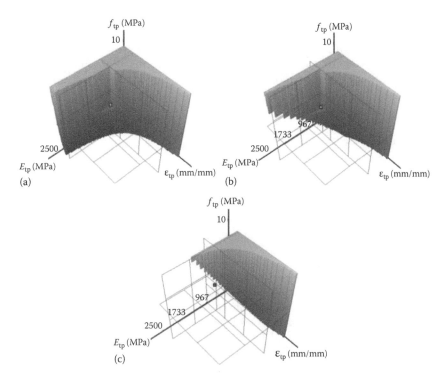

Figure 5.8 Examples of compatibility subspaces determined with the ANCOMP code taking into account the modulus of elasticity E_{tp}, the elongation at break, ε_{tp}, and the tensile strength f_{tp} showing the influence of crack width wd, in the concrete substrate: (a) w_d = 0.1 mm on cracked substrate; (b) w_d = 0.1 mm on non-cracked substrate (compatibility); and (c) w_d = 0.2 mm on noncracked substrate (lack of compatibility); maximal crack width variation Δw = 0.03 mm in all cases.

5.8 CONCLUSION

When compatibility issues are properly addressed in repair systems, durability of the bond can be achieved, ensuring a lasting coexistence of the repair material and substrate concrete. Incompatibility issues are the primary cause for premature debonding and repair failures. Development of reliable guidelines addressing compatibility with special emphasis on the factors related to dimensional compatibility issues is needed for the repair industry to evolve as an engineering discipline.

The results clearly demonstrate usability of the compatibility model for prediction of behavior of the system: concrete substrate–repair material during service and an indication of the good cooperation that will not be met either at the stage of selection of repair and protective coating, or during the use of the building structure. Modeling tools such as ANCOMP show very interesting potential to implement successfully the compatibility approach in the design of concrete repairs and to assist the development of innovative durable surface repair and treatments for concrete. Efforts are needed toward the development of applications that will take into account all compatibility considerations and adapted characterization test protocols.

REFERENCES

Bissonnette, B. and Pigeon, M. (1995) Tensile creep at early ages of ordinary, silica fume and fiber reinforced concretes. *Cement and Concrete Research*, **25**(5), 1075–1085.

Bissonnette, B. and Pigeon, M. (March 2000) Le comportement viscoélastique du béton en traction et la compatibilité déformationnelle des réparations. *Materials and Structures*, **33**(226), 108–118.

Courard, L. and Bissonnette, B. (2008) Compatibility performance as a fundamental requirement for the repair of concrete structures with self-compacting repair mortars. *RILEM PRO 54: Proceedings of the Fifth International Symposium on Self-Compacting Concrete*, Ghent, Belgium, Vol. 1, pp. 667–675.

Cusson, D. (1995) Durable concrete patches. *Construction Canada*, **37**(4), 34–39.

Czarnecki, L., Clifton, J.R., and Głodkowska, W. (1992) Problem of compatibility of polymer mortars and cement concrete system. *International Colloquium Materials and Restoration*, Esslingen, Germany, pp. 964–971.

Czarnecki, L., Garbacz, A., Łukowski, P., and Clifton, J.R. (1999) Polymer composites for repairing of portland cement concrete: Compatibility project. NISTIR 6394, NIST, Gaithersburg, MD, 87p.

Czarnecki, L., Głodkowska, W., and Piatek, Z. (2004) Estimation of compatibility of polymer i polymer-cement composites with ordinary concrete under short-time load conditions. *Archives of Civil Engineering*, **L**(1), 133–150.

Czarnecki, L. and Runkiewicz, M. (2005) On the compatibility measure in the repair systems. In: *Proceedings of the International Conference on Concrete Repair, Rehabilitation and Retrofitting (ICCRRR'2005)* (Eds. H. Beushausen, F. Dehn, and M.G. Alexander), Balkema, Rotterdam, the Netherlands, pp. 1003–1008.

Dehawah, H.A.F., Basanbul, I.A., Maslehuddin, M., Al-Sulaimani, G.J., and Baluch, M.H. (March-April 1994) Durability performance of repaired reinforced concrete beams. *ACI Materials Journal*, **91**(2), 167–172.

Emberson, N.K. and Mays, G.C. (1990a) Design of patch repairs: Measurements of physical and mechanical properties of repair systems for satisfactory structural performance. *International Conference on Protection of Concrete*, Dundee, U.K., pp. 937–954.

Emberson, N.K. and Mays, G.C. (1990b) Significance of property mismatch in the patch repair of structural concrete—Part 1: Properties of repair systems. *Magazine of Concrete Research*, **42**(152), 147–160.

Emberson, N.K. and Mays, G.C. (1990c) Significance of property mismatch in the patch repair of structural concrete—Part 2: Axially loaded reinforced concrete members. *Magazine of Concrete Research*, **42**(152), 161–170.

Emmons, P.H. and Vaysburd, A.M. (1993a) Compatibility considerations for durable concrete repairs. *Transportation Research Record*, **1382**, 13–19.

Emmons, P.H. and Vaysburd, A.M. (1993b) Factors affecting durability of concrete repair—The contractor's viewpoint. *Proceedings of the Fifth International Conference on Structural Faults and Repair*, Edinburgh, U.K., pp. 253–268.

Emmons, P.H. and Vaysburd, A.M. (1994) Factors affecting durability of concrete repair—The contractor's viewpoint. *Construction and Building Materials*, 8(1), 5–16.

Emmons, P.H. and Vaysburd, A.M. (April 1995) Performance criteria for concrete repair materials. Phase I, Technical Report REMR-CS-47, U.S. Army Corps of Engineers, Waterways Experiment Station, Vicksburg, MS, 113p.

Emmons, P.H. and Vaysburd, A.M. (March 1997) Corrosion protection in concrete repair: Myth and reality. *Concrete International*, **19**(3), 47–56.

Emmons, P.H., Vaysburd, A.M., McDonald, J.E., Poston R.W., and Kesner K.E. (2000) Selecting durable repair materials: Performance criteria. *Concrete International*, 3, 38–45.

Felt, E.J. (1956) Resurfacing and patching concrete pavement with bonded concrete. *Proceedings of the Highway Research Board*, **35**, National Research Council, Washington, DC, pp. 444–469.

Forgey, C. (June 2005) Changing the color of concrete. *Concrete International*, 27(6), 78–81.

Garbacz, A. (2013) Importance of concrete substrate preparation for effective repair. *Construction Materials*, 9, 10–13 (in Polish).

Garbacz, A. and Głodkowska, W. (2011) Compatibility as an important factor affecting repair durability. *Proceedings of the European Symposium on Polymers in Sustainable Construction (ESPSC'2011)*, September 6–7, Warsaw, Poland, pp. 105–106.

Garbacz, A. and Głodkowska, W. (2012) Compatibility of Elastic PCC Coating and Concrete Substrate, *International Journal for Restoration of Buildings and Monuments*, 18(3/4): 229–238.

Garbacz, A., Głodkowska, W., Courard, L., Bissonnette, B., and Vaysburd, A.M. (2014) Patch repair: Compatibility issues. *Proceedings of the Conference Concrete Solutions*, September 1–3, Belfast, U.K., pp. 231–236.

Głodkowska, W. (1994) Compatibility of polymer composite—Cement concrete system. Ph.D. Dissertation, Warsaw University of Technology, Warsaw, Poland (in Polish).

Głodkowska, W. (2003) *Fundamentals and Method of Selection of Properties of Composites for the Repair and Protection of Concrete.* Koszalin University of Technology, Koszalin, Poland (in Polish).

Głodkowska, W. (2011) Forecasting crack resistance of short- and long-term loaded coatings. *ACME Journal*, 10(1), 33–44.

Głodkowska, W. and Staszewski, M. (2007) Weatherability of coating materials for protection of concrete. In *Adhesion in Interfaces of Building Materials: A Multi-Scale Approach. Advances in Materials Science and Restoration* (Eds. L. Czarnecki and A. Garbacz), AMSR No. 2. Aedificatio Publishers, Freiburg, Germany, 193–205.

Gu, P., Beaudoin, J.J., Tumidajski, P.J., and Mailvaganam, N.P. (1997) Electrochemical incompatibility of patches in reinforced concrete. *Concrete International*, 19(8), 68–72.

Gulikers, J.J.W. and Van Mier, J.G.M. (1991) The effect of patch repairs on the corrosion of steel reinforcement in concrete. *Second CANMET/ACI International Conference on Durability of Concrete*, Supplementary Papers, Montreal, Quebec, Canada, pp. 445–460.

Hanley, B., Coverdale, T., Miltenberger, M.A., and Nmai, C.K. (May/June 1998) Electrochemical compatibility in concrete repair. *Concrete Repair Bulletin*, 11(3), 12–15.

ICRI Guideline No. 03733 (January 1996) Guide for selecting and specifying materials for repair of concrete surfaces. International Concrete Repair Institute (ICRI), Des Plaines, IL, p. 34.

Marosszeky, M. (1992) Stress performance in concrete repairs. *International Conference on Rehabilitation of Concrete Structures*, Melbourne, Victoria, Australia, pp. 467–474.

Marosszeky, M. and Wang, D. (1991) The study of the effect of various factors on subsequent corrosion. *Second CANMET/ACI International Conference on Durability of Concrete*, Supplementary Papers, Montreal, Quebec, Canada, pp. 585–606.

Marosszeky, M., Yu, J.G., and Ng, C.M. (1991) Study of bond in concrete repairs. ACI Special Publication SP-126-70, Vol. 2, American Concrete Institute, Detroit, MI, pp. 1331–1354.

Morgan, D.R. (1996) Compatibility of concrete repair materials and systems. *Construction and Building Materials*, 10(1), 57–67.

Neville, A.M. (1995) *Properties of Concrete*, 4th edn. Essex, England: Longham.

Pigeon, M. and Bissonnette, B. (1999) Tensile creep and cracking potential of bonded concrete repairs. *Concrete International*, 21(11), 31–35.

Pigeon, M. and Saucier, F. (1992) Durability of repaired concrete structures. In: *Advances in Concrete Technology* (Ed. V.M. Malhotra), CANMET, pp. 741–773.

Plum, D.R. (1990) The behavior of polymer materials on concrete repair and factors influencing selection. *The Structural Engineer*, 68(17), 337–345.

RILEM TC 124-SRC Draft Recommendation (1994) Repair strategies for concrete structures damaged by steel corrosion. *Materials and Structures*, 27, 415–436.

Saucier, F. and Pigeon, M. (1991) Durability of new-to-old concrete bondings. ACI Special Publication SP-128-43, Vol. 1, American Concrete Institute, Detroit, MI, pp. 689–705.

Schrader, E. and Kaden, R. (1987) Durability of shotcrete, ACI SP-100. *Katherine and Bryant Mather International Conference on Durability of Concrete*, Atlanta, GA, Vol. 2, pp. 1071–1101.

Schrader, E.K. (1992a) Mistakes, misconceptions, and controversial issues concerning concrete and concrete repairs—Part I. *Concrete International*, 14(9), 52–56.

Schrader, E.K. (1992b) Mistakes, misconceptions, and controversial issues concerning concrete and concrete repairs—Part II, *Concrete International*, 14(10), 48–52.

Schrader, E.K. (1992c) Mistakes, misconceptions, and controversial issues concerning concrete and concrete repairs—Part III. *Concrete International*, 14(11), 54–59.

Tabor, L.J. (1985) Twixt old and new: Achieving a bond when casting fresh concrete against hardened concrete. *Second International Conference on Structural Faults and Repair*, London, U.K., pp. 57–63.

Vaysburd, A. et al., *Conc. Int.*, 22(12), 39, 2000.

Vaysburd, A.M. (2005) Repairing concrete: Random thought on concrete ills and treatment prescriptions. *International Journal of Materials and Product Technology*, 23(3/4), 164–176.

Vaysburd, A.M. (2006) Holistic system approach to design and implementation of concrete repairs. *Cement and Concrete Composites*, 28, 671–678.

Vaysburd, A.M., Bissonnette, B., and Brown, C.D. (2009) Compatibility and concrete repair. *The Construction Specifier*, 62(1), 44–53.

Vaysburd, A.M., Emmons, P.H., McDonald, J.E., Poston, R.W., and Kesner, K.E. (2000) Selecting durable repair materials: Performance criteria—Field studies. *Concrete International*, 22(12), 39–45.

Yeoh, K.M., Cleland, D.J., and Long, A.E. (1992) The effect of environmental conditions on interface adhesion properties of concrete patch repairs. *Second International Conference on Inspection, Appraisal, Repairs and Maintenance of Building and Structures*, Jakarta, Indonesia, pp. 237–243.

Chapter 6

Surface preparation

6.1 OBJECTIVES

According to the basic principle of concrete structure repair, formalized in many standards and guidelines, existing concrete surfaces must be treated to remove carbonated or unsound material and roughen their profile and thereby promote good bond with repair materials. The efficiency of repair, restoration, or, more generally, connections depends primarily on the quality of the surface (or surfaces) and therefore the pretreatment.

The surface preparation of concrete substrate is one of the four general requirements for repair and protection presented in the European Standard EN 1504-10 (2003) *Products and systems for the protection and repair of concrete structures—Definitions—Requirements—Quality control and evaluation of conformity. Site application of products and systems and quality control of the works.* The general requirements for all methods of repair are

- Weak, damaged, and deteriorated concrete and, when necessary, sound concrete shall be removed.
- Cleaning shall be carried out after roughening or concrete removal (water-based methods make this unnecessary).
- Microcracked or delaminated concrete, including that caused by the technique of cleaning, roughening, or removal, which reduces bond or structural integrity to an unacceptable extent, shall be subsequently removed or remedied.
- The finished surface shall be visually inspected and tested by tapping with a hammer to detect loose concrete.

The substrate quality is largely influenced by the surface preparation technique used. Selection of the most suitable technique(s) depends on the nature and objectives of the repair project, (Table 6.1).

More detailed guidelines for selection of surface preparation techniques were prepared in the framework of the REHABCON "Strategy for maintenance and rehabilitation in concrete structures" (2000). It describes successive steps in repair process; the removal and roughening methods (Table 6.2)

Table 6.1 Goals of surface preparation of concrete substrate according to EN 1504

| Preparation techniques | | Methods of repair | | | | | |
Goal	Description	Impregnation	Surface coating	Filling cracks, voids, or interstices	Application of mortar and concrete	Installing bonded rebars in holes	Plate bonding
Concrete removal	• Removal shall be kept to a minimum. • Removal shall not reduce structural integrity beyond the structure's ability to perform its function (temporary support). • Depth of carbonation and the concentration profiles of chloride or other contamination shall be established and taken to account. • The methods are: mechanical, thermal, and water blasting (hydrodemolition ≥ 60 MPa).				X	X	X
Roughening	• The texture of roughed surface shall be appropriate for the products applied and shall be specified. • Roughening is used for removal of concrete up to 15 mm depth. • Gives a textured surface with good bonding when a new layer of concrete is cast or sprayed onto the old one. • The methods are: mechanical, percussion and abrasion, grit and sandblasting, and water blasting (≤60 MPa).				X		X
Cleaning	• The substrate shall be free from dust, loose material, surface contamination, and materials which may reduce bond by affecting adhesion of the repair material and/or its penetration into the porosity of the near-to-surface layer surface. • The cleaned substrate shall be protected from further contamination if cleaning is not carried out immediately before application. • The methods are: mechanical, percussion and abrasion, grit and sandblasting, and water blasting (≤18 MPa).	X	X	X	X	X	X

Table 6.2 Concrete removal and roughening methods recommended in REHABCON (2000)

Category	Description	Comments
Crushing	Uses hydraulically powered jaws to crush and remove concrete.	• Appropriate for large, horizontal surfaces. • Bulky machinery.
Cutting	(Water jet cutting, diamond blade cutting, stitch drilling, etc.) Empty full-perimeter cuts to disjoint concrete for removal as a unit.	• Expensive. • Bulky machinery.
Milling	(Hydromilling, rotary head milling, scarification) Abrasion or cavitation erosion techniques to remove concrete (Figure 6.9).	• Appropriate for large surfaces. • Bulky machinery. • Generates lots of dust.
Needle gun	Metal needles strike perpendicularly the surface at high speed (3003–6000 impacts per minute), removing the weakened concrete (Figure 6.4).	• Only useful for flat (even) or almost flat (even) surfaces.
Sandblasting (dry sand)	A sand jet impacts the substrate, removing both dirt and deteriorated concrete, but not to a large extent.	• Mostly ineffective for concrete removal. • Requires experienced staff. • Dusty method.
Hydrode-molition	Water jet at a very high pressures demolishes the concrete.	• Transmits few vibrations to the concrete. • Requires few manual labor. • Trouble with slurry management. • Bulky machinery. • Can be ineffective if the concrete contains extremely coarse aggregate.
Flame removing	Removes layers of concrete not bigger than 5 mm due to thermal shock (Figure 6.13).	• It must be followed by another surface treatment. • Dangerous.
Impacting	(Hammers, pneumatic hammers, bush hammers, scabbling) Repeated striking of the surface with the hammer, a chisel, or/and a pneumatic hammer to fracture and remove the concrete.	• Pneumatic hammers should not be so powerful that they microcrack the surrounding sound concrete. • After it is done, it is good to brush the surface with a wire brush. • Cannot work closer than 10 cm to the reinforcements. • Pneumatic bush hammers are faster but less effective (excellent finishing).
Pre-splitting	(Chemical expansive agents) Employ wedging forces in a designed pattern of boreholes to produce a controlled cracking of the concrete to facilitate removal.	• Expensive. • Ineffective on reinforced concrete.

are followed by superficial cleaning (Table 6.3). The purpose of cleaning methods is to improve the bond between the substrate and the repair material applied by eliminating loose particles and remove the stains that are on the surface of concrete. Dust and loose fine material existing on the substrate may contain enough unhydrated cement to set in the presence of moisture, and it has to be removed before setting could occur.

Table 6.3 Concrete superficial cleaning methods recommended in REHABCON (2000)

Preparation	Description	Comments
Water washing	Washing should be done from the top of the structure down ward. If water alone is not enough to clean the concrete, the following materials can be added: a mild soap, a stronger one, ammonia, or vinegar.	A mist spray is recommended.
Steam cleaning	Steam is good for removing dirt of atmospheric origin.	Expensive.
High-pressure water jetting	Pressure of 150–1000 atmospheres. Tests are recommended to determine the optimal pressure. Water can be heated or include additives to maximize the effects.	Requires experienced stuff. Water must be very clean. Incompatible with epoxy mortars. Transmit no vibrations to the concrete. Fast and not expensive.
Sandblasting (dry sand)	A sand jet impacts the substrate, removing both dirt and deteriorated concrete, but not to a large extent.	Opens the pore structure of the concrete. Requires experienced stuff. Generates lots of dust.
Sandblasting (wet sand)	The same as the dry technique but by wetting the sand with water before it leaves the pipe.	Much less dust. Slower than the dry one. Lot of cleanup needed.
Flame cleaning	Will remove organic materials that do not respond to solvents (Figure 6.13).	Can cause scaling of the concrete surface and may produce fumes.
Mechanical cleaning	Some tools like grinders, buffers, or brushes may be required to remove the more stubborn stains from concrete.	Can remove more concrete than is desirable.
Chemical cleaning	Organic solvents can be used with little dissolution. Inorganic solvents such as ammonium hydroxide can be purchased in ready-mixed solutions. Acid-based cleaners soften the stains making them easier to remove with a brush.	Manufacturer's recommendations about use and safety must be carefully followed.

Concrete removal and surface preparation techniques can be classified on the base of their operating modes as follows:

- *Dynamic techniques*: they take advantage of the low resistance to the impact of the concrete. The concrete is broken by applying repeated shocks.
- *Mechanical techniques*: they exploit the low tensile strength of concrete. It creates a stress higher than the tensile strength.
- *Abrasive techniques*: wearing of the surface by friction with hard instruments or particles.
- *Thermal techniques*: the concrete is molten like lava or slag and thermal dilatation gradients are induced.
- *Chemical techniques*: breaking up concrete by dissolving some of its phases or constituents (e.g., acid-etching).

Concrete removal and surface preparation techniques may be also classified depending on the effect they produce as follows:

- Preparation techniques in depth: disintegration and concrete demolition on its full thickness (dynamite and explosives). They are used when you want to completely demolish any structure or part of a structure.
- Preparation techniques on the surface:
  - Removal of deteriorated concrete: e.g., jackhammer, high-pressure water jetting.
  - Preparation of the surface to be repaired: e.g., cleaning, sandblasting.
- Techniques of cutting and drilling: cutting the various components of the structure by sawing or by running a series of closely spaced perforations. These techniques are often used to clearly delimit the extent of some repairs.

The performance and effectiveness of the various concrete removal and preparation techniques can vary greatly depending on the quality of concrete. A combination of two or more techniques may be necessary to perform surface preparation or repair of a surface. The selection of the concrete removal techniques depends on many factors such as:

- Site location (indoor, outdoor) and special conditions of accessibility and environment
- Restrictions on noise, dust, vibrations, fumes, debris collection, or liquid waste
- Scaffolding
- Thickness of demolition and type of defect elimination
- Quality of the concrete
- Presence of reinforcement
- Type of structure (deformable or not)

As the main concern of this book is surface engineering, we only developed the description of techniques used for concrete removal and surface preparation.

6.2 CONCRETE REMOVAL TECHNIQUES

For concrete removal, the methods by percussion are the most frequently used. These tools are often used because of ease of use and relatively low cost. But there are also techniques based on the projection of solids or liquids.

6.2.1 Hammering

The "tap" consists of the percussion of concrete by means of a tip (Figure 6.1). Different types of stitching may be distinguished:

- *Handmade*
  - Limited to small horizontal, vertical, and ceiling surfaces and deformable mass or support
  - Performed at the point or chisel
  - Irregular and slow process (artisanal)
  - Mainly used for the removal of surface deposits or opening cracks
- *Mechanical*: Pneumatic or hydraulic hammers are to be used with caution. It is best to use power tools and only as a last resort if no other technique is applicable. They are suitable for removing heavy deposits on the massive concrete. Light breakers are generally characterized with a weight less than 14 kg and 1000–2000 rounds/min. Performance varies greatly depending on the quality of concrete, the presence of reinforcement, and the operator.
- *Heavier breakers* (Figure 6.2) will use a hammer weighing between 15 and 30 kg. They require vertical punching and will induce more vibrations and lower quality of surface (cracking).

Figure 6.1 Schematic representation of concrete removal with a light handheld breaker (*bush hammer*) (From Emmons, P.H., *Concrete Repair and Maintain Illustrated*, RSMeans Company, Kingston, MA, 1994, 295p.) and a typically resulting substrate profile.

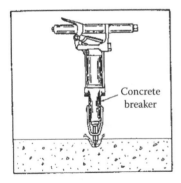

Figure 6.2 Schematic representation of concrete removal with a heavy handheld breaker (*jackhammer*) (From Pigeon, M. and Beaupré, D., *Guide pratique des techniques de démolition et d'extraction du béton détérioré*, Association des constructeurs de routes et grands travaux du Québec [ACRGTQ], Montreal, Quebec, Canada, 1990, 35p.) and an example of surface preparation.

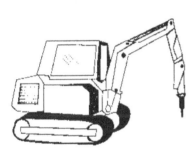

Figure 6.3 Schematic representation of a machinery-mounted breaker (From Emmons, P.H., *Concrete Repair and Maintain Illustrated*, RSMeans Company, Kingston, MA, 1994, 295p.) and an example of surface preparation.

- *Machinery-mounted breaker*: This equipment can be installed on a metal mini-arm (Kubota type). These devices have a higher rate of production, and they can be used in any position (Figure 6.3). They cause a lot of noise and vibration. The resulting surface is usually of very poor quality (cracking), with a high risk of damaging the rebar.
- *Hammer needle*: The pounding of the surface is by means of needles (Figure 6.4). The yield depends on the number of needles and the incident energy. It is only suitable for small areas. This is a slow process that can be also applied to steel surfaces.

6.2.2 Scabbling

Scabbling is a mechanical process intended to remove a thin layer of concrete from a structure, typically achieved by compressed-air-powered machines. A typical scabbling machine uses several heads, each with several

Figure 6.4 Concrete surface preparation with needle gun. (From Anonymous, http://www.hss.com/hire/p/power-scraper-needle-hilti, accessed on March 29, 2015.)

carbide or steel tips that peck at the concrete. It operates by pounding a number of tipped rods down onto the concrete surface in rapid succession (Figure 6.5). It takes several passes with the machine to achieve the desired depth. (Figure 6.5). The degree of fineness of processing is proportional to the number of peaks. It is like jackhammering, which means a violent process. It applies only to concrete surfaces and can only be used in the case of mass support (horizontal surfaces, vertical surfaces, or ceilings). Distinction can be made between

- Manual hammering:
  - Slow process
  - Eliminates thicknesses or crusts
  - Applied to remove old coatings
  - Causes the breakup of a few millimeters thick skin surface
- Mechanical roughening:
  - Very energetic process that can disrupt the surface structure
  - Can only be used with extreme caution in prestressed concrete
  - Best suited for corrections flatness
  - Preferably use electrical tools, the incident energy is lower than with pneumatic tools
  - Applied to large horizontal surfaces

Figure 6.5 Concrete surface preparation by scabbling: (a) general review, (b) scabbling head, and (c) scabbled surface (From authors).

6.2.3 High-pressure water jetting

High-pressure water jetting technique uses one or more jets of water at high pressure (138–276 MPa) to break or remove deteriorated concrete surface. The hydraulic pressure may be adjusted depending on the depth or demolition according to a criterion based on the minimum quality of the concrete to be removed. Deep demolitions (complete) may be achieved in the case of thin, reinforced concrete slabs (100 mm or more). Hydrodemolition is a technique of selective demolition.

There are devices with one spray lances that can be used by an operator (Figure 6.6). This type of equipment is used for hard to reach places. There are also facilities with several jets that may be used on essentially horizontal surfaces (Figure 6.7). These advanced systems with multiple jets whose movement can be programmed are used to accurately determine the depth and area of concrete to be removed.

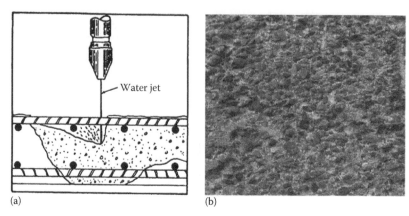

(a)   (b)

Figure 6.6 (a) Principle of high-pressure water jetting with a single jet (From Pigeon, M.
and Beaupré, D., *Guide pratique des techniques de démolition et d'extraction du
béton détérioré*, Association des constructeurs de routes et grands travaux du
Québec [ACRGTQ], Montreal, Quebec, Canada, 1990, 35p.) and (b) example
of the surface after preparation. (Adapted from Courard, L. et al., Effect of
concrete substrate texture on the adhesion properties of PCC repair mortar,
Proceedings of the International Symposium on Polymers in Concrete, Guimareas,
Portugal, 2006, pp. 99–110.)

Figure 6.7 Water jetting machine with multiple jets. (From Anonymous, http://www.
hydro-demolition.org.uk/images/access-all-1.jpg, accessed on March 28, 2015.)

Figure 6.8 State of the concrete and reinforcement after demolition with high-pressure water jetting. (From Anonymous, http://www.lcwhitford.com/images/ upload/PhotoGallery-GA_Bridges/GDOT, Hydro-demolitionConcreteOverlay, August 2014, resized.jpg, accessed on March 28, 2015.)

Some advantages and disadvantages of the technique are:

- A lot of noise.
- Low vibration.
- Little dust.
- A lot of liquid residues that can clog drainage systems. There should be regular cleaning of pipes and surfaces.
- Very good surface quality.
- Very high efficiency (does the same as 20–40 men with jackhammers).
- Relatively high cost.
- Carefully clean the rebar (Figure 6.8).
- Requires large amounts of clean water (100–250 L/min).

It is necessary in any case to properly adjust the pressure versus the quality of the concrete. This technique will also reveal the differences in compactness or homogeneity of the concrete.

6.2.4 Scarification and milling

Scarification is a technique used to quickly remove a thickness of several inches of concrete for a new resurfacing or profile correction. The scarifier is provided with a plurality of rotating metal spikes (jagged rolls), which collide and break the concrete (Figure 6.9). This process is also referred to as milling operation. Some large facilities are equipped with vacuums

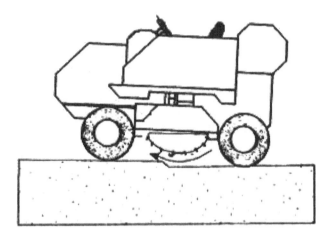

Figure 6.9 Scheme scarification operations. (From Emmons, P.H., *Concrete Repair and Maintain Illustrated*, RSMeans Company, Kingston, MA, 1994, 295p.)

Figure 6.10 Portable scarifier. (From Bissonnette, B. et al., *Concr. Int.*, 28(12), 49, 2006.)

or recyclers to collect particles (Figure 6.10). Scarification cannot be used when there is a risk of encountering steel reinforcement.

It causes a fast enough surface smoothness and does not treat the hollow; little dust is produced. This is a very fast method (100 m²/h) and provides a pretty good surface preparation (fine texture).

6.3 SURFACE PREPARATION TECHNIQUES

Brushing, sandblasting, jet beads, and grinding are techniques that allow the removal of a thin layer (a few mm) of concrete.

6.3.1 Manual brushing (wire brush)

Brushing with a wire brush is effective in removing poorly adherent deposits on the concrete and cleaning the exposed rebar. However, its action is limited to small areas.

6.3.2 Mechanical brushing (rotary wire brushes)

Mechanical brushing eliminates slightly sticky deposits. However, it generates a very smooth surface, and the dust produced is incorporated into the pores of the concrete. This technique is limited to a few specific cases where other methods are not applicable.

6.3.3 Sandblasting

Cleaning by sandblasting consists of a stream of sand particles driven at high speed by a jet of compressed air (Figure 6.11). Shocks wear out the concrete and/or clean the surface. It allows the removal of laitance, dirt, oils, etc. This technique generates a lot of dust.

Sandblasting is one of the best methods that gives a very good surface roughness and is ideal for gluing reinforcement. It can only be performed by skilled labor.

6.3.4 Hydroblasting

The sandblasting technique combines the projection under high pressure of water and sand. It has the disadvantages of being less effective than

Figure 6.11 Resurfacing concrete with sandblasting and an example of a surface after preparation. (From Harman, T. and Kojundic, T., Crystalline Silica Rule under review at Office of Management and Budget, *Concrete Bridge Views* (US Dept. of Transportation, Federal Highway Admistration), 70 (on-line www.concretebridgeviews.com/i70/article3, 2013 and from the authors.)

sandblasting and delivers a moist surface, which requires drying. However, it will prevent dust and sparks.

6.3.5 Shotblasting

Bead blasting (shotblasting) consists of projecting high-speed steel beads (2 mm diameter) (Figure 6.12). The steel balls are drawn immediately after their projection on the concrete to be reused. This technique produces less dust. The depth of the concrete which is removed is typically a few millimeters but may be greater if desired.

The three processes (sanding, sandblasting, and shotblasting) are applied to concrete, on massive or deformable surfaces as well as large, medium, or small surfaces.

6.3.6 Grinding

Grinding is a technique which removes a certain thickness by abrasion of the concrete. These are usually juxtaposed diamond discs that abrade the concrete up to the desired depth. This technique enables removing bituminous or rubber membranes where sandblasting and bead blasting are almost ineffective.

It forms very fine dust that fill the pores and capillaries and must be followed by a strong cleaning (sometimes difficult to conduct properly). It applies primarily as a pretreatment of local unevenness.

6.3.7 Thermal methods

The treatment of the surface flame is never used as the sole treatment. It is intended to burn certain materials as a deposit before mechanical

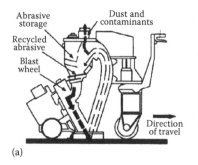
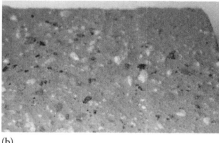

(a)   (b)

Figure 6.12 (a) Principle of operation of a unit bead blasting (shotblasting) (From Mailvaganam, N.P., *Repair and Protection of Concrete Structures*, CRC Press, Boca Raton, FL, 1992, 473p.) and (b) an example of surface after preparation (From authors).

Figure 6.13 Oxyacetylene burner. (From Anonymous, http://en.informatiitehnice.com/wp-content/uploads/2014/01/Flame-Spraying.jpg, accessed on March 27, 2015.)

processing. We must never proceed with the torch for removing hydrocarbon deposits because there is a risk of migration of these substances deeper into the concrete.

Oxyacetylene flame causes a thermal shock to ±3200°C to obtain a concrete adapted to receive a layer of leveling compound. Most frequently, the apparatus consists of a "rake" consisting of nozzles arranged along a rail perpendicular to the direction of movement (Figure 6.13).

The technique requires a skilled workforce, knowing to adjust the burner and knowing the optimum travel speeds. It provides a dry surface. Repeated running should be avoided because of the risk of cracking, popping, or explosion of aggregates. It can cause excessive decarbonation.

Surfaces contaminated with hydrocarbons require an update preliminary adjustment and it should particularly pay attention to the steel reinforcement near the surface.

Concrete surfaces will be afterward treated intensively with the rotary wire brush directly after heat treatment, preferably by milling or scarification.

6.3.8 Chemical methods

These techniques will be applied with caution to avoid any infiltration into the joints by controlling the duration of treatment and the depth of penetration of corrosive materials. Rinsing and pH control are essential. These methods are never applied to prestressed structures. Methods using acids are the most common:

- *HCl (3%–5% by weight)*: Fast and easy process, provides a good surface, dangerous for reinforcements, take all precautions to protect personnel.
- H_3PO_4 *(10%–15%)*: Safer than HCl for reinforcements but ineffective vs. fats and oils.

Methods with basic products are used most often with surfactants or emulsifiers to remove oily, fatty deposits. This is the best method for removing animal fats.

Solvents are used for degreasing steel. The solvent must be completely removed from the substrate before applying the resin. Steam may be used to enhance the action of alkalis and surfactants.

6.4 EFFECTS OF PREPARATION TECHNIQUES ON SURFACE ROUGHNESS

As a result of surface preparation, different types of roughness are obtained. The European Standard EN 1504-10 does not define requirements for optimal roughness level. Mechanical interlocking and contact angle modifications are two fundamental effects of surface "roughness." The first one is in relation to the "waving" of the surface, while the second is more affected by "micro-roughness" (Czarnecki et al., 2003): the value of the contact angle made by the liquid on the solid surface is modified by the roughness, according to the Wenzel relation (see Chapter 4). Table 6.4 presents the effect of various surface preparation techniques on surface topography of the C20/25 concrete substrate. The obtained profile amplitude parameters clearly confirmed that as the surface preparation "aggressiveness" increases, the surface roughness increases too (Garbacz et al., 2005). Abbott's curves (Figure 6.14) showed that the surfaces prepared by grinding, sandblasting, and hand milling have similar geometry like the formworked concrete surface. The surfaces resulting from shotblasting (20 and 35 s, respectively) and after mechanical milling belong to the second group of surface geometry with medium roughness. A significantly rougher surface with large peaks and holes was obtained after shotblasting for 45 s. The results showed that the time of shotblasting or the use of hand or mechanical milling does not change the amplitude parameters of the profile significantly. This behavior was already observed for the comparison between sandblasted and polished concrete surfaces (Courard, 1998).

Amplitude parameters are able to quantitatively characterize the profile by analyzing holes, peaks, frequencies, and amplitudes of the irregularities. The difference between the profiles is more effective at the level of waviness than roughness (Figure 6.15); on its own waviness profile, roughness amplitude is not statistically different for the different profiles (Garbacz et al., 2005).

Table 6.4 Selected parameters of surface geometry of concrete substrate subjected to various surface preparation techniques

View of the profile on cross section (mag. 10×)	3D visualization with surfometry	Roughness parameters
(a) Without treatment (NT)		
		SRI = 242 mm W_a = 5 μm W_t = 39 μm W_P = 13 μm C_F = 9 μm C_R = 6 μm
(b) Grinding (GR)		
		SRI = 0.72 mm W_a = 32 μm W_t = 219 μm W_P = 111 μm C_F = 55 μm C_R = 57 μm

(Continued)

Table 6.4 (Continued) Selected parameters of surface geometry of concrete substrate subjected to various surface preparation techniques

View of the profile on cross section (mag. 10×)	*3D visualization with surfometry*	*Roughness parameters*
(c) Sandblasting (SB)		$SRI = 1.40$ mm $W_a = 49$ µm $W_t = 434$ µm $W_p = 117$ µm $C_F = 77$ µm $C_R = 50$ µm
(d) Shotblasting 35 s (SHB35)		$SRI = 1.59$ mm $W_a = 215$ µm $W_t = 1086$ µm $W_p = 570$ µm $C_F = 406$ µm $C_R = 289$ µm
(e) Mechanical milling (MM)		$SRI = 1.05$ mm $W_a = 179$ µm $W_t = 867$ µm $W_p = 448$ µm $C_F = 351$ µm $C_R = 188$ µm

Source: Adapted from Garbacz, A. et al., *Mag. Concr. Res.*, 57(1), 49, 2005.

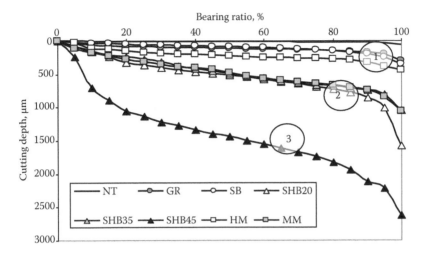

Figure 6.14 Abbott's curve for different surface preparation techniques applied to C20/25 concrete substrates: grinding (GR), sandblasting (SB), shotblasting (treatment time: 20 s for SHB20, 35 s for SHB35, and 45 s for SHB45), milling (hand—HMIL and mechanical—MMIL); NT—substrate without preparation. (From Garbacz, A. et al., *Mag. Concr. Res.*, 57(1), 49, 2005.)

6.5 MICROCRACKING AND BRUISING

The main problems arise from *co-lateral effects* of the treatment, especially due to microcracks parallel to the surface (Bissonnette et al., 2006). Superficial cracking, often referred to as "bruising," is considered as one of the most important parameters influencing adhesion in a repair system and should be taken into account when the preparation method is selected. Figure 6.16 presents the examples of the view of surfaces of C20/25 concrete slabs subjected to the most common mechanical surface preparation techniques (Garbacz et al., 2005) very often used in industrial floor industry.

Based on microscopic observations of samples taken at the surface of the slabs subjected to sandblasting, scarifying, high-pressure water jetting, and jackhammering, Bissonnette et al. (2006) showed that impact hammers induce significant damage in the prepared substrate, while the other techniques leave a much sounder surface (Figure 6.17).

The number of cracks and the total crack length resulting from the preparation with jackhammer were observed to be significantly higher than with any of the other investigated techniques (Figure 6.18). Moreover, it could be seen that increasing the jackhammer weight—and thus, the impact energy—causes the extent of cracking (length and number of cracks) to increase significantly. In addition, it should be recalled that handheld breaker treatments was followed by sandblasting in the experiments reported here; had

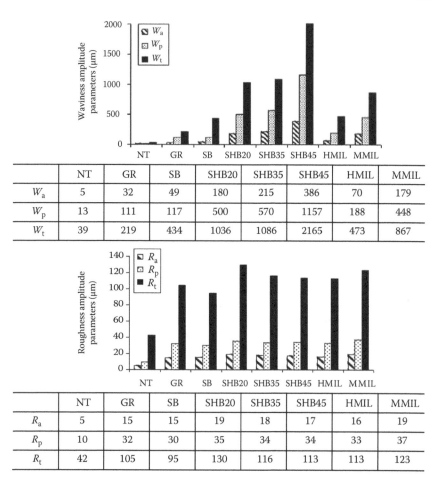

	NT	GR	SB	SHB20	SHB35	SHB45	HMIL	MMIL
W_a	5	32	49	180	215	386	70	179
W_p	13	111	117	500	570	1157	188	448
W_t	39	219	434	1036	1086	2165	473	867

	NT	GR	SB	SHB20	SHB35	SHB45	HMIL	MMIL
R_a	5	15	15	19	18	17	16	19
R_p	10	32	30	35	34	34	33	37
R_t	42	105	95	130	116	113	113	123

Figure 6.15 Waviness (up) and roughness (bottom) parameters after application of different surface preparation techniques for C20/25 concrete substrate (symbols as in Figure 6.14). (Adapted from Garbacz, A. et al., *Mag. Concr. Res.*, 57(1), 49, 2005.)

the sandblasting operation not been performed, the results in terms of cracking would presumably have been worse.

Investigations for concrete substrates of different compressive strength subjected to polishing, hydroblasting, scabbling, and hydrodemolition (Courard et al., 2014) confirmed that more aggressive surface treatment techniques greatly influence microcracking. The density of microcracks L_A was two times higher after jackhammering and, to a lesser extent, hydrodemolition than after dry sandblasting and polishing (Table 6.5). These findings are consistent with those of many authors (Cleland et al., 1992; Schrader, 1992).

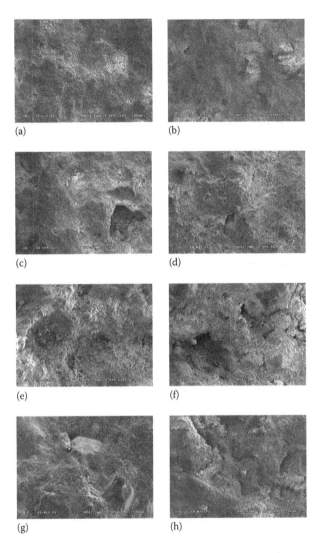

(a)

(b)

(c)

(d)

(e)

(f)

(g)

(h)

Figure 6.16 Surface of C20/25 concrete substrate without surface treatment (a) and after different mechanical preparations: (b) grinding, (c) sandblasting, (d) shotblasting—20 s, (e) shotblasting—35 s, (f) shotblasting—45 s, (g) hand milling, (h) mechanical milling; SEM; magn. 100×. (From Garbacz, A. et al., *Mag. Concr. Res.*, 57(1), 49, 2005.)

It appears from these results that techniques such as sandblasting, scarifying, and water jetting are preferable to the use of breakers as they clearly leave fewer defects in the substrate, thus promoting the development of better bond characteristics. The respective influence of the various surface preparation techniques on microcracking is summarized in Table 6.6.

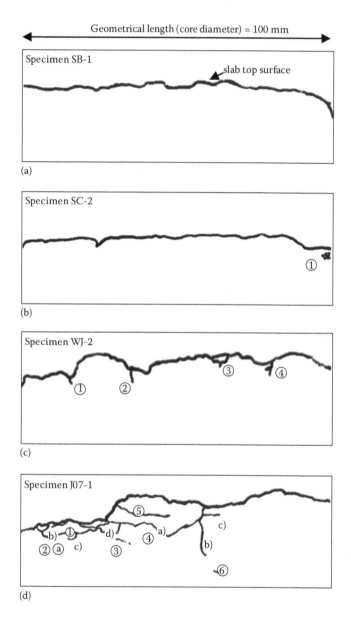

Figure 6.17 Typical superficial cracking induced by different preparation techniques, as observed on the cross section of sawn 100 mm cores observed under a microscope: (a) sandblasting (SB), (b) scarifying (SC), (c) water jetting (WJ), (d) 7 kg jackhammering + sandblasting (J07).   *(Continued)*

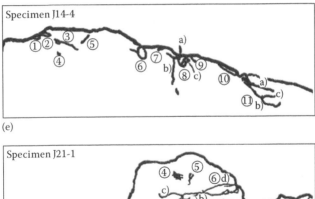

(e)

(f)

Figure 6.17 (Continued) Typical superficial cracking induced by different preparation techniques, as observed on the cross section of sawn 100 mm cores
observed under a microscope: (a) sandblasting (SB), (b) scarifying
(SC), (c) water jetting (WJ), (d) 7 kg jackhammering + sandblasting
(J07), (e) 14 kg jackhammering + sandblasting (J14), and (f) 21 kg
jackhammering + sandblasting (J12). (From Bissonnette, B. et al.,
Concr. Int., 28(12), 49, 2006.)

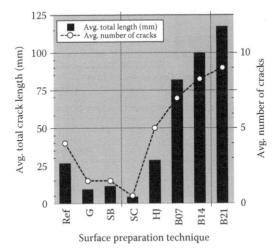

Figure 6.18 Total crack length and number of cracks yielded by different surface preparation techniques (note: Ref, no treatment; G, grinding; SB, sandblasting;
SC, scarifying; WJ, water jetting; B07, 7 kg handheld breaker + sandblasting;
B14, 14 kg handheld breaker + sandblasting; B21, 21 kg hand-held breaker +
sandblasting). (From Bissonnette, B. et al., *Concr. Int.*, 28(12), 49, 2006.)

Table 6.5 Density of microcracks (L_A) for different compressive strength concrete substrates subjected to polishing, hydroblasting, scabbling, and hydrodemolition

Image of cross section of C40 sample	Sample	L_A (mm/cm²)
Polishing		
	C30/37	1.52
	C40/50	3.23
	C50/60	1.55
Wet sandblasting (hydroblasting)		
	C30/37	2.09
	C40/50	3.45
	C50/60	1.96
Scabbling		
	C30/37	4.34
	C40/50	3.89
	C50/60	5.19
Hydrodemolition (very high-pressure water jetting)		
	C30/37	3.96
	C40/50	4.91
	C50/60	4.05

Source: Courard, L. et al., *Cement Concr. Compos.*, 46, 73, 2014.

Table 6.6 Microcracking risk level for different surface preparation techniques

Surface preparation	Microcracking risk		
	Very low	Moderate	High
Abrasive (sand) blasting	X		
Steel shotblasting	X		
Scarifying		X	
Needle scaling		X	
High and ultra high-pressure water jetting	X		
Scabbling			X
Milling/rotomilling			X
Flame blasting		X	

Source: Bissonnette, B. et al., Bonded cement-based material overlays for the repair, the lining or the strengthening of slabs and pavements, RILEM STAR Report Volume 3, 193-RLS RILEM TC, Springer, Dordrecht, the Netherlands, 2011, 175p.

6.6 COMPARISON AND LIMITATION OF TECHNIQUES

This is not only necessary to improve rehabilitation and repair work specifications but is also a fundamental issue in quantitative evaluation through an appropriate field test and, eventually, suitable acceptance criteria for concrete surface preparation work. Table 6.7 shows all the techniques commonly available in the market and their effects and consequences on the environment.

6.7 MOISTENING THE SURFACE

In the EN 1504-10 (2003), it is stressed that, before and during application of the products and systems, the temperature and moisture content of the substrate and the environmental conditions such as the temperature, the relative humidity, the dew point, the rate of moisture changes as influenced by rain and wind shall be considered. It was shown in Chapter 4 that the adhesion between the repair material and concrete substrate depends on moisture condition of the substrate prior to application of repair materials. The moisture condition of the substrate will determine the rate of movement of water from the repair mortar to the substrate concrete due to the moisture imbalance between the two layers. During the process of water movement, two phenomena can be observed: penetration of water from the repair mortar in the capillaries of substrate concrete and hydration of cement paste in the repair. Depending on the type of product which is used for repair, water may act in a positive or negative way.

With cementitious repair products or systems, the use of bonding agents can reduce bonding if the bond coat sets before the application of subsequent products, notably due to high water absorption by the dry concrete substrate. Where cementitious products or systems are used without a bond agent, the concrete substrate shall be well pre-wetted but free from water on the surface at the time of the application. The condition of the substrate shall be specified where a bond coat is used.

Sprayed concrete and sprayed mortar used as repair material shall comply with the standard for sprayed concrete (EN 14487-1 (2005) and EN 14487-2 (2006)). The need for pre-wetting of the substrate shall be considered. It depends upon the condition of its application and the composition of the products and systems used. Spray fog deposits or overspray shall be removed from surrounding areas and from the substrate before sprayed concrete or mortar is applied.

In the case of application of coatings, the maximum and minimum temperature and moisture content of the substrate and the environment will be specified and selected as appropriate to the surface coating, hydrophobic impregnation, or impregnation material. In the case of polymer coating, the moisture content should not exceed 4% (RILEM, 2011).

According to RILEM TC 193-RLS (Bissonnette et al., 2011), the optimum water condition of a concrete substrate for a particular cement-based

Table 6.7 Surface preparation techniques commonly available in the market and their effects and consequences on the environment

Removal method	Principle behavior	Depth action (mm)	Important advantages	Important disadvantages
Sandblasting	Blasting with sands	No	No microcracking	Not selective, leaves considerable sand.
Scrabbling	Pneumatically driven bits impact the surface	No (6)	No microcracking, no dust	Not selective.
Shotblasting	Blasting with steel balls	No (12)	No microcracking, no dust	Not selective.
Grinding (planning)	Grinding with rotating lamella	No (12)	Removes uneven parts	Dust development, not selective.
Flame cleaning	Thermal lance	No	Effective against pollutions and painting, useful in industrial and nuclear facilities	The reinforcement may be damaged, smoke and gas development, safety considerations limit use, not selective.
Milling (scarifying)	Longitudinal tracks are introduced by rotating metal lamellas	Yes (75)	Suitable for large volume work, good bond if followed by water flushing	Microcracking is likely, reinforcement may be damaged, dust development, noisy, not selective.
Pneumatic (jack) hammers (chipping), handheld or boom-mounted	Compressed-air-operated chipping	Yes	Simple and flexible use, large ones are effective	Microcracking, damages reinforcement, poor working environment, slow production rate, not selective.
Explosive blasting	Controlled blasting using small, densely spaced blasting charges	Yes	Effective for large removal volumes	Difficult to limit to solely damaged concrete, safety and environmental regulations limit use, not selective.
Water jetting (hydrodemolition)	High-pressure water jet from a unit with a movable nozzle	Yes	Effective (especially on horizontal surfaces), selective, does not damage reinforcement or concrete, improved working environment	Water handling, removal in frost degrees, costs for establishment.

Source: Bissonnette, B. et al., Bonded cement-based material overlays for the repair, the lining or the strengthening of slabs and pavements, RILEM STAR Report Volume 3, 193-RLS RILEM TC, Springer, Dordrecht, the Netherlands, 2011, 175p.

repair material can be determined by preliminary testing, using different moisture surface conditioning:

- Saturated Surface Dry (SSD)
- Saturated Surface Wet (SSW)
- Unsaturated Surface Dry (USD)
- Unsaturated Surface Wet (USW)

In cases when such testing cannot be performed, SSD moisture conditioning should be applied. Under this condition, the substrate looks damp but contains no free water on the surface. The surface absorbed all the moisture possible but does not contribute water to the repair material mixture.

6.8 CONCLUSION

Surface preparation and cleaning operations are critical in order to remove all deteriorated concrete and contaminants that may adversely influence bond and to promote maximum penetration of the repair material or surface treatment into the concrete substrate. This is essential for developing the interlocking and adhesion forces needed for a strong, lasting bond. A wide range of methods and techniques are available to accomplish the work, from light surface roughening to large-volume concrete removal. While many of them can yield a sound surface with adequate characteristics for intended applications, some others may induce significant damage in the prepared concrete surface. For instance, commonly used tools such as concrete breakers, which are widely available and offer attractive production rates, produce harmful bruising in the substrate. In addition to the other technical and economical considerations, this has to be taken into account when specifying or selecting the surface preparation technique.

REFERENCES

Bissonnette, B., Courard, L., Fowler, D.W., and Granju, J.L. (2011) Bonded cement-based material overlays for the repair, the lining or the strengthening of slabs and pavements. RILEM STAR Report Volume 3, 193-RLS RILEM TC, Springer, Dordrecht, the Netherlands, 175p.

Bissonnette, B., Courard, L., Vaysburd, A., and Bélair, N. (2006) Concrete removal techniques: Influence on residual cracking and bond strength. *Concrete International*, 28(12), 49–55.

Cleland, D.J., Yeoh, K.M., and Long, A.E. (1992) The influence of surface preparation methods on the adhesion strength of patch repairs for concrete. *Proceedings of the Third Colloquium on Materials Science and Restoration*, Esslingen, Germany, pp. 858–871.

Courard, L. (1998) Parametric definition of sandblasted and polished concrete surfaces. *Proceedings of the 9th International Congress on Polymers in Concrete (ICPIC)*, Bologna, Italy (Ed. P. Sandrolini), Edizioni Casma s.r.l., Bologna pp. 771–778.

Courard, L., Piotrowski, T., and Garbacz, A. (2014) Near-to-surface properties affecting bond strength in concrete repair. *Cement & Concrete Composites*, 46, 73–80.

Courard, L., Schwall, D., Garbacz, A., and Piotrowski, T. (2006) Effect of concrete substrate texture on the adhesion properties of PCC repair mortar. *Proceedings of the International Symposium on Polymers in Concrete*, Guimareas, Portugal, pp. 99–110.

Czarnecki, L., Garbacz, A., and Kostana, K. (2003) The effect of concrete surface roughness on adhesion in industrial floor systems. *Proceedings of the Fifth International Colloquium on Industrial Floors*, January 21–23, Esslingen, Germany, pp. 168–174.

Emmons, P.H. (1994) *Concrete Repair and Maintain Illustrated*. Kingston, MA: RSMeans Company, 295p.

EN 1504-10 (2003) Products and systems for the protection and repair of concrete structures—Definitions—Requirements—Quality control and evaluation of conformity. Site application of products and systems and quality control of the works CEN, Bruxelles, Belgium.

EN 14487-1 (2005) Sprayed concrete. Definitions, specifications and conformity CEN, Bruxelles, Belgium.

EN 14487-2 (2006) Sprayed concrete. Execution CEN, Bruxelles, Belgium.

Garbacz, A., Gorka, M., and Courard, L. (2005) Effect of concrete surface treatment on adhesion in repair systems. *Magazine of Concrete Research*, 57(1), 49–60.

Harman, T. and Kojundic, T. (2013) Crystalline Silica Rule under review at Office of Management and Budget. *Concrete Bridge Views*. (US Dept. of Transportation, Federal Highway Admistration), 70 (on-line www.concrete-bridgeviews.com/i70/article3).

Mailvaganam, N.P. (1992) *Repair and Protection of Concrete Structures*. Boca Raton, FL: CRC Press, 473p.

Pigeon, M. and Beaupré, D. (1990) *Guide pratique des techniques de démolition et d'extraction du béton détérioré*. Montreal, Quebec, Canada: Association des constructeurs de routes et grands travaux du Québec (ACRGTQ), 35p.

REHABCON IPS-2000-00063 (2000) Annex H "Patching" EC DG ENTR-C-2 Innovation and SME Program "Strategy for maintenance and rehabilitation in concrete structures." 37p.

RILEM (2011) Bonded cement-based material overlays for the repair, the lining or the strengthening of slabs and pavements. (Eds. B. Bissonnette, L. Courard, D.W. Fowler, and J.L. Granju, Rilem Star Report Volume 3, Springer, U.K., 177pp.

Schrader, E.K. (1992) Mistakes, misconceptions, and controversial issues concerning concrete and concrete repairs (part 3). *Concrete International*, 14(11), 54–59.

Chapter 7

Surface treatment of concrete and adherence

7.1 TYPES OF SURFACE TREATMENTS FOR CONCRETE

A variety of surface treatments can be performed to repair and/or protect concrete elements. These surface treatments can be classified in accordance with the following four categories:

- Surface protection
- Crack repair
- Surface repair/overlay
- Strengthening

Most materials and material categories available for the first three types of treatments are summarized in Tables 7.1 through 7.3, respectively, together with their main characteristics, advantages, and disadvantages. In the case of surface-applied strengthening, linings (sheets, strips, etc.) made with different materials (steel, carbon fibers embedded in a resin and a variety of other fiber reinforced polymers [FRP]) are typically installed in selected areas of the surface using high-performance adhesives (epoxy-type, polyurethane-type, etc.).

The effectiveness and durability of many of these treatments depend primarily on their ability to adhere strongly and durably to the concrete surface onto which they are applied. In the specific cases of penetrating sealers (silane, siloxane) and concrete cracking repair materials, bonding with the concrete develops respectively within the surface pores and within the surface cracks of the treated substrate. In any case, the interaction between the treatment material and the treated concrete surface determines the quality and mechanical strength of the bond.

Table 7.1 Materials for concrete surface protection

Category	Generic	Breathability (water vapor transmission)
Penetrating sealer/ hydrophobic treatment	Silane	✓
	Siloxane	✓
Sealer/impregnation[a]	Acrylic	✓ (low)
	Epoxy	✓ (low)
	Linseed oil	✓ (low)
Coatings	Polymer-modified cementitious	✓
	Acrylic	✓
	Methacrylate	✓
Membranes	Liquid-applied (polymer-modified asphalt, polyurethane, etc.)	
	Sheet-applied (rubberized asphalt, expandable bentonite adhered to high density polyethylene, etc.)	
	Built-up systems (staggering of reinforcing fabric and resin layers)	

Source: American Concrete Institute, ACI 546.3 R-14—Guide to materials selection for concrete repair, American Concrete Institute, Farmington Hills, MI, June 2014, 76pp.

[a] Some paints can be classified in that category.

Table 7.2 Materials for concrete cracking repair

| Generic | Crack movement | | Purpose of restoration | |
	Dormant	Active	Watertightness	Structural
Cement grout	✓		✓	
Polymer grout	✓		✓	
Polymer-modified (cementitious) grout	✓		✓	
Epoxy resin	✓			✓
Polyurethane grout	✓	✓	✓	
Flexible epoxy resin or polyurea	✓	✓	✓	
Polymer sealant[a]	✓	✓	✓	
Strip-and-seal systems	✓	✓	✓	
Methacrylate (HMWM or MM)[b]	✓			✓

Source: American Concrete Institute, ACI 546.3 R-14—Guide to materials selection for concrete repair, June 2014, American Concrete Institute, Farmington Hills, MI, 76pp.

[a] Polyurethane, silicone, polyether, polysulfide, etc.
[b] High molecular weight methacrylate (HMWM), methyl methacrylate (MM).

Table 7.3 Materials for concrete surface repair

Generic	Material grade		Applications			
			Deep repairs		Shallow	
	Concrete	Mortar	Horizontal surfaces	Vertical/overhead surfaces	repairs	Overlays
Ordinary Portland cement (OPCᵃ)	●		✓	✓	✓	✓✓
High performance (HPᵇ)	●	●	✓	✓✓✓		✓✓
Self-compacting (SC)	●	●	✓	✓✓✓		✓✓
Polymer-modified (cementitious)	●	●	✓	✓✓✓	✓	✓✓
Polymer	●	●	✓	✓✓✓	✓	✓✓
Other specialty bindersᶜ		●	✓	✓	✓	✓✓

ᵃ Often incorporate supplementary-cementing materials (SF, fly ash, slag).
ᵇ Typically incorporate silica fume.
ᶜ Calcium aluminate cement, magnesium ammonia phosphate cement, etc.

7.2 EVALUATION OF ADHERENCE

7.2.1 Bond strength

Bond strength, or adhesion, relates to the ability of the two materials to act as one. This section addresses the mechanical bond between the repair material and substrate concrete and its quantitative evaluation.

A repair material should have sufficient bond strength to the substrate concrete such that it does not separate from the substrate concrete. It is desirable for the bond strength to be larger than the minimum requirements so that any reductions due to field-application conditions are not as critical. Bond strengths that exceed the tensile strength of either the repair material or the substrate will cause the failure to occur in the weaker material if sufficient interface stresses result from shrinkage, thermal movement, or other factors.

7.2.1.1 Repair and overlay materials

The schematics in Figure 7.1 show a range of existing types of tests for evaluating the interface bond strength between a repair material or treatment and a concrete substrate. A number of these methods have been standardized by different agencies, as listed in Table 7.4.

In these test methods, bond strength is usually inferred from the recorded direct tension, slant shear, direct shear, or torsional shear test values, because failure rarely occurs at the bond line. Failures that occur away from the bond line imply that the bond strength is greater than the failure load in the test. When failure occurs at the bond line in direct tension tests, the measured tensile force is the actual bond strength.

For the evaluation of tensile bond strength, the most widely used method is the pull-off test (CAN/CSA A23.2-6B; EN 1542; ASTM C1583; ICRI No. 210.3). The test method consists of drilling a core through the repair material down to a minimum depth within the substrate, gluing a steel dolly onto the top of the core with epoxy, and to pull on the steel dolly using a special loading rig. The tensile bond strength is equal to the maximum recorded stress when failure occurs in the interfacial zone, whereas a lower boundary value of bond strength is obtained when failure occurs elsewhere. It is generally considered to be more appropriate for repair applications, because it involves a single mode of stress and direct tension at the bond interface. The CAN/CSA A23.2-6B, EN 1542, ASTM C1583, and ICRI No. 210.3 test procedures are suitable for both field and laboratory evaluations. The main parameters influencing pull-off test results are the transfer plate (dolly) thickness and diameter, the core drilling depth, the speed of loading, the adhesive type and thickness, and the number of tests.

Even with its limitations, pull-off test remains the most common way to estimate the quality of bond in repairs. In addition to the determination of bond strength, it can provide useful information on failure.

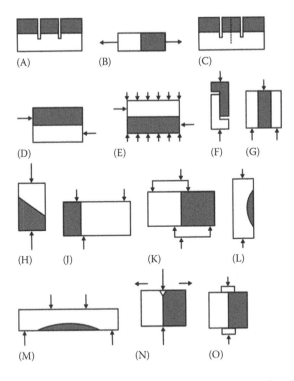

Figure 7.1 Schematics of various test methods to determine interface bond strength
(Rilem TC-RLS, 2011); (A) pull-off test, (B) uniaxial tensile test, (C) interface
shear strength, (D) shear bond test, (E, F, G) indirect shear bond test, (H) slant
shear test, (J) guillotine test, (K) interface shear test, (L, M) interface shear
and tensile bond test, (N) tensile interface strength in bending, (O) interface tensile strength with splitting test. (From Bissonnette, B. et al., Bonded
cement-based material overlays for the repair, the lining or the strengthening
of slabs and pavements, RILEM STAR Report Volume 3, 193-RLS RILEM TC,
Springer, Dordrecht, the Netherlands, 2011, 175p.)

Slant shear bond tests (ASTM C882) measure the resistance to sliding
between the repair material and concrete substrate along an inclined surface. The interface is subjected to combined shear and compressive stress.
The measured bond strength may be influenced by the compressive strength
of the substrate concrete and the roughness of the bonding surface. The
Michigan Department of Transportation direct shear bonding test measures the maximum shearing force applied along the interface between the
repair material and sawn substrate concrete. In the torsional shear test procedure, a ring glued to the surface is twisted off using a torque housing with
eccentric loading. The housing is anchored to the surface and the loading
is typically performed with the same pulling unit as in the pull-off test
procedure (different adapters). The peak force is recorded and is converted
to a shear stress.

Table 7.4 Methods and standards for evaluating bond strength

Category	Standard method
Direct tensile bond	ACI 503R
	ASTM C882
	ASTM C1404
	ASTM C1583
	CAN/CSA A23.2-6B
	EN 1542
	EN 12636
	ICRI 03739
Direct shear bond	Michigan Department of Transportation Direct Shear Bonding Test
Slant shear bond	ASTM C882
	ASTM C1042

7.2.1.2 Traffic-bearing elastomeric coatings and membranes

The following test methods can be used to measure adhesion of traffic-bearing elastomeric coatings and membranes to concrete (ACI 546.3 R-14):

- ASTM D4541
- ASTM C794, as modified by ASTM C957
- ASTM D903

ASTM D4541 involves adhering a dolly or stud to the coating surface and using a portable adhesion tester to measure the tension pull-off. The test identifies the weakest part of the coating system, either intralayer cohesion, interlayer adhesion, bond to the substrate, or a weak substrate. The test may be performed in the laboratory or in the field as a quality control measure. It is a destructive test that requires patching in the field. ASTM C794 modified by ASTM C957 and ASTM D903 are peel-off tests. In the ASTM C794 procedure, a piece of cloth is embedded in the elastomeric coating, adhered to a substrate, and pulled off the substrate. For ASTM D903, a flexible material is bonded to a flexible or rigid substrate material with the elastomeric coating, and then the flexible material is pulled off the substrate material. The numerical results obtained from each of these procedures are for comparison with the results for other materials tested by the same procedure.

7.2.2 Assessment of bond quality

The durability of repair directly depends on bond quality. For a given combination of treatment material and existing concrete and thus specific interface composition and microstructural characteristics, bond quality actually refers to the intimate and uniform contact between the treatment applied on the substrate, as well as to the absence of defects. Even if bond strength determined with a limited number of pull-off tests appears satisfactory, interfacial defects may still be present in some areas. Local defects such as contaminants, voids, cracks, delaminations, and others can be highly detrimental for the durability of the bond in promoting the accumulation of moisture and eventually triggering debonding.

7.2.2.1 Microscopical examinations

Bond characteristics between concrete and a surface treatment can be appraised visually through microscopical observations, the scale of interest depending on the type of treatment and information looked after. For example, the photographs of Figure 7.2 illustrate some of the differences found at the interface between a slurry (used as a bonding agent) and an

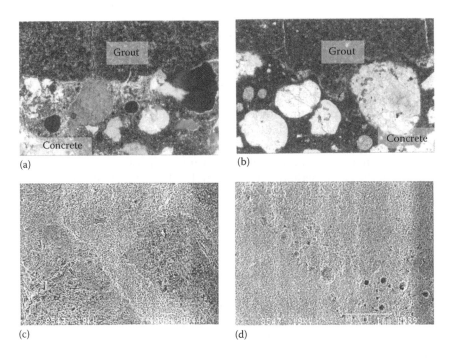

Figure 7.2 Comparative influence of the moisture condition of the concrete substrate on the interface characteristics of a slurry used as a bonding agent (prior to the placement of a repair material): (a) interfacial area of a slurry placed on a wet (sandblasted) substrate (underside view), (b) interfacial area of a slurry placed on a dry (ground) substrate (underside view), (c) surface of contact of a slurry layer (bottom) placed on a saturated ground substrate, and (d) surface of contact of a slurry layer (bottom) placed on a dry (ground) substrate. (From Courard, L., *Mag. Concr. Res.*, 57(5), 273, 2005.)

existing concrete substrate, depending on the moisture condition of the latter. The moisture condition at the time of material placement visibly can affect the porosity and entrapment of air bubbles among other things.

Such observations have contributed, for instance, to explain why in the case of some polymer-modified mortars (PCC), optimal tensile bond strength results were yielded when saturation levels of the concrete substrate ranged between 55 and 80, with or without slurry (Courard and Darimont, 1998).

7.2.2.2 Nondestructive testing for the assessment of bond quality

Due to their invasive character, the use of the various bond strength tests described in previous section for quality control or condition assessment in repaired structures is often restricted by owners and managers. The adaptation of nondestructive testing (NDT) methods/protocols for the assessment of bond quality and uniformity thus represents a quite promising alternative.

In the recent years, a handful of NDT methods (see Table 3.8) have been developed for the evaluation of concrete structures (Malhotra and Carino, 2004; Hoła et al., 2015). Contrary to traditional methods involving core testing, NDT techniques can yield information about the material properties without affecting the integrity and serviceability of the elements tested in a structure. In addition to their nondestructive character, the main advantages of NDT methods are the rapidity of use, the possibility of on-site evaluation, and the monitoring capability (repeated measurements performed at given locations of the structure in service).

Most of the NDT methods referred to in EN 1504-10 and in the Concrete Repair Manual (ACI/ICRI, 2013) for the assessment of bond quality are based on propagation of stress waves, notably the ultrasonic echo method (U-E), the ultrasonic pulse velocity method (UPV) and the impact-echo method (I-E). (see Table 3.8). So far, the assessment has generally been based on the detection of internal anomalies, which interfere with the propagation of certain type of waves. By monitoring the response of the tested element when it is subjected to such waves, the presence of the anomaly can be inferred. Recently, the ground penetrating radar (GPR) method has been increasingly used for similar purposes.

Repair systems are difficult to appraise with NDT methods, because of the many factors influencing the stress wave propagation. Taking into account the classification proposed by Adams and Drinkwater (1997), two main types of defects can occur in these systems that can affect stress wave propagation:

- Adhesion type (in the repair–substrate interfacial area): various types of *non-zero-air volume* debonding (e.g., interfacial voids, delamination) and *zero-air volume* debonding—weak adhesion areas (e.g., due to the presence of dust, oil, etc.)

- Cohesion type (in repair material or/and concrete substrate): porosity, cracks, honeycombing, partially nonhardened resin in the case of polymer materials

To select the appropriate NDT method for bond quality assessment, the following factors should be taken into account (Adams and Drinkwater, 1997; Malhotra and Carino, 2004):

- Type and size of defects that may be expected in the investigated area
- Repair thickness
- Type of repair material (cement-based or polymer composites [PCs])
- Characteristics of the concrete substrate (roughness, microcracking, saturation level)

The first two factors depend mainly on the NDT method and device being used. For example, in the case of ultrasonic testing the transducers of low frequency are preferable and in the case of impact-echo (IE) testing, the depth and size of the detectable defects depend on the selected impactor diameter.

It should be stressed that of all the efforts devoted toward the development of NDT techniques to assist the condition evaluation of concrete structures, very little were actually dedicated on the quantitative evaluation of bond quality (e.g., Czarnecki et al., 2006; Santos et al., 2011; Sadowski and Hoła, 2014; Garbacz, 2015).

7.2.2.3 Multilayer repaired structure

In the case of multilayer systems such as a repaired concrete element, the propagation of stress waves depends on the differences in acoustic impedance between the repair material and the concrete substrate (Malhotra and Carino, 2004). For two dissimilar materials, the incident wave energy (A_i) is partially transmitted (A_{tr}) when it crosses their interface, as a fraction (A_r) of it is reflected. The wave reflection is characterized by the coefficient of reflection (R), which is determined as follows:

$$R = \frac{A_r}{A_i} = \frac{Z_2 - Z_1}{Z_2 + Z_1} \tag{7.1}$$

where $Z_1, Z_2 = c_p \times d_v$, acoustic impedance of materials 1 and 2 (kg/m^2 s); c_p is the P-wave velocity (m/s); and d_v is the volume density (kg/m^3).

The coefficient of reflection at the concrete–air interface is nearly equal to 1, which means there is almost complete reflection at the interface. This is why the NDT methods based on the stress wave propagation are useful for detecting *non-zero*-volume defects such as air voids and delamination.

Using finite-element method (FEM) modeling, Garbacz and Kwaśniewski (2006) carried out a comprehensive numerical study to evaluate the potential ability of stress wave propagation for assessing repair bond with the impact-echo method. They selected to study how the presence of a polymer influences stress wave propagation in polymer–cement composites (PCCs) and PCs, these repair materials being among those exhibiting the most significant differences (properties/characteristics) with ordinary Portland cement (OPC) concrete in practice. The simulations were performed for three selected repair materials having different elastic moduli and three different values of the R coefficient (0.08, 0.18 and 0.35). The clear frequency peak corresponding to reflection from the interface was observed only for $R = +0.35$. Comparable results were obtained by Sansalone and Lin (1994a,b). FEM simulations and experimental investigations with the I-E method have shown that the presence of an interface is usually detectable, even in the case of high adhesion, provided that the absolute value of the R coefficient is larger than 0.24.

The effect of polymer content on stress wave propagation in PCC repair systems was analyzed by Garbacz (2015). The values of acoustic impedance

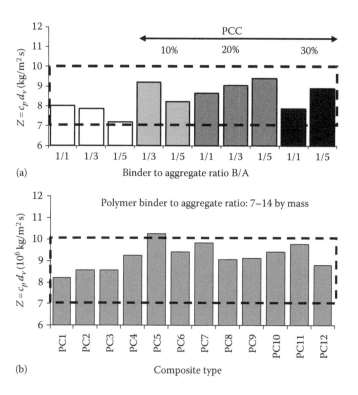

Figure 7.3 Estimation of acoustic impedance of: (a) polymer–cement (PCC) and (b) polymer (PC) mortars and concretes with varying polymer contents using ultrasonic method. (From Garbacz, A., *Bull. Pol. Acad. Sci.: Tech. Sci.*, 63(1), 77, 2015.)

for PCC were estimated taking into account the measured densities and pulse velocities. The addition of polymer was found to increase the acoustic impedance by approximately 5%–15% in comparison to that recorded for the reference mixture (Figure 7.3a). These values are close to the typical range (Lewińska-Romicka, 2000) for OPC concretes and mortars (dotted line in Figure 7.3). The value of the reflection coefficient for the PCC–OPC concrete interface can be estimated to be less than +0.15.

A similar analysis was performed for repair polymer (PC) mortars and concretes (w/o Portland cement), taking into account acoustic impedance values that were calculated based upon pulse velocity and density data obtained by Garbacz and Garboczi (2003) for mixtures prepared using different polymers (epoxy and vinylester), a constant binder content, and varying aggregate proportions. In general, the acoustic impedance values of the PC composites are higher than those found for PCC composites (Figure 7.3b). The absolute value of reflection coefficient for the PC–OPC concrete interface was in turn estimated to be less than +0.18.

The low values of reflection coefficient for both types of polymer repair materials indicate that a large fraction of the stress wave energy will be transmitted through the interface between the repair layer and the OPC concrete substrate. The detection of flaws at the PCC–OPC interface can thus be performed in accordance with procedures developed for monolithic concrete elements or structures.

7.2.2.4 Detection of defects at the interface (non-zero-air volume defects)

Defects located at the interface, for example delaminations, air voids, etc., are detrimental to the adhesion of repairs and surface treatments. The presence of interfacial defects can result from many different causes associated with surface preparation, materials characteristics and preparation, placement and curing conditions, service loads, exposure conditions and/or the ageing effects. From an NDT prospective, such defects may be classified as non-zero air volume defects. Owing to the presence of air voids, these types of defects should in principle be more easily detectable, but the substrate roughness and microcraking resulting from the surface preparation operation may affect the elastic wave propagation through the interface. Beside, one of the main challenges with the use of NDT is to extract from the recorded signals quantitative information on adhesion.

The potential of I-E and UP-E methods for the detection of non-zero-volume defects at the repair interface was investigated by Garbacz et al. (2005). Concrete (C20/25) test slabs were subjected to different types of surface preparation prior to repair, such as to generate a wide range of roughness and microcracking levels. Surface roughness characteristics were evaluated through high frequency filtration of the surface profile determined with a mechanical profilometer on each prepared slab (Garbacz et al., 2005). The slabs were then overlaid (thickness, 10 mm) with a proprietary polymer–cement repair mortar. On half of the test slabs, a bonding agent was applied prior to the repair mortar placement, as suggested by the material's manufacturer.

After proper conditioning, measurements with impact-echo and utrasonic pulse echo NDT methods were carried out on the overlaid test slabs and bond strength between the repair material and the concrete substrate was evaluated with pull-off testing (acc. to EN 1542). In the end, samples were taken from the different slab series and the interface characteristics were examined using light microscope.

In the case of slabs repaired without the use of a bonding agent, an increasing volume of air voids was found at the interface as the surface roughness of the prepared substrate increased, and the corresponding pull-off strength values were observed to be lower. In the other test slabs, the bonding agent helped preventing the entrapment of interfacial voids and the influence of roughness upon bond strength was observed to be relatively limited.

Correspondingly, the I-E test results (Garbacz, 2010) show a statistically significant relationship between pull-off strength results and the P-wave velocity for the repair system without bonding agent, while it does not exhibit any correlation in the case of repairs with the bonding agent (Figure 7.4a). It can be further seen in Figure 7.4b that in the absence of a bonding agent, the P-wave velocity increases with the substrate roughness, because of the increasing amount of interfacial voids. No such relationship is observed for the repair system incorporating a bonding agent, which can be explained by the fact that the bonding agent filled more effectively the superficial irregularities in the repaired concrete substrate. Overall, the results obtained with the I-E technique tend to indicate that the P-wave propagation through the repair system is not altered as such by the tortuosity of the bond line but rather by the presence of interfacial voids.

The same repair systems were tested further with the ultrasonic pulse echo method using digital ultrasonic flaw detector and a pair of transducers having a nominal frequency of 500 kHz. Each received A-scan consisted of characteristic peaks corresponding to the reflection from the interface. Analysis of the relationship between the amplitude of maximum frequency peaks and the pull-off strength on Figure 7.4c reveals the following overall trend: as the pull-off strength increases, the amplitude of the highest peak recorded decreases. As can be observed on the graph, that trend characterizes essentially the response obtained for the repair series without a bonding agent. In the case of the latter, a statistically significant relationship was found between the average surface profile roughness and the amplitude value of the highest peak recorded (Figure 7.4d), again because the volume fraction of air voids in the interfacial area increased with the surface roughness of the prepared substrate.

The assessment of bond quality with the I-E method was carried out for higher strength concrete repairs (Garbacz et al., 2006). A series of C40/50 test slabs were prepared using four different techniques (sandblasting, scabbling, high-pressure hydro jetting, and grinding) and overlaid with a proprietary

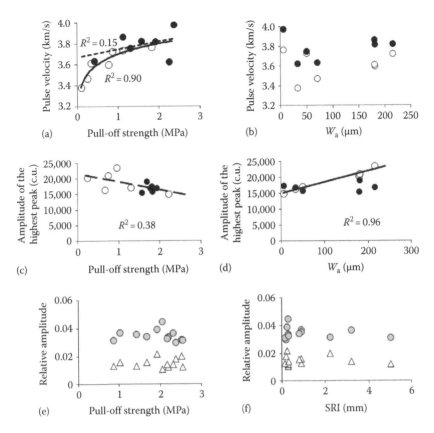

Figure 7.4 NDT survey of C20/25 repaired concrete slabs (with and without a bonding agent) with I-E and UP-E test methods—Relationships between stress wave propagation parameters and repair system characteristics: I-E pulse velocity vs. (a) pull-off strength; and (b) mean waviness of profile, W_a; amplitude of the highest peak of frequency spectrum of UP-E signal (c.u.—conventional unit) vs. (c) pull-off strength; and (d) mean waviness of profile, W_a overlays with [●] and without [○] the bond coat); amplitude of bottom (○) and interface (Δ) I-E frequency peaks versus (e) pull-off strength and (f) SRI. (From Garbacz, A., *Bull. Pol. Acad. Sci.: Tech. Sci.*, 63(1), 77, 2015.)

self-compacting PCC mortar (30 mm thick). For these repair systems, two specific frequency spectrum ranges were analyzed from the recorded I-E signals: bottom peak and interfacial peak. For both relative amplitude parameters analyzed, either as a function of pull-off strength or the level of roughness of the substrate, none of the relationships was found to be statistically significant (Figure 7.4e and f). This indicates that the peak amplitude is not a meaningful parameter for the estimation of bond strength in repair systems.

To gain a better understanding of the effect of interfacial air voids upon stress wave propagation, FEM calculations (Kwaśniewski and Garbacz, 2008) were conducted assuming two different scenarios: repair of the

C40/50 concrete substrate roughness where profile valleys either enclosing air voids or completely filled by the repair material system. The surface roughness characteristics corresponded to those determined experimentally (Figure 7.5a) for real slabs after sandblasting and hydro jetting respectively. The simulation results are consistent with the experimental results, confirming further that the presence of large air voids at the interface can significantly influence the stress wave propagation (Figure 7.5b and c). If the surface profile irregularities in the substrate are adequately filled, the surface roughness does not significantly influence the resulting frequency spectrum (Figure 7.5d). Similar results were obtained by Santos et al. (2011), although their FEM simulations showed that the pulse decreases in the presence of rough interfaces, due to greater wave dispersion.

The investigations of different repair systems placed on a variety of concrete substrates (concrete grade, surface preparation) showed that for both the I-E and ultrasonic methods, the roughness and microcracking of the concrete substrate do not significantly affect the stress wave propagation through the repair system, provided that the bond quality is sufficient (absence of large voids at the interface). However, parameters describing roughness and microcracking of the concrete substrate can still be important for improving the reliability of the bond strength evaluation using more complex signal analyses (e.g., wavelet approach, artificial neural networks of stress waves resulting from NDT methods). Examples of such analyses were provided by Sadowski and Hoła (2014), who showed that substrate roughness is an important factor for the prediction of bond strength between the concrete layers in concrete floors using the nondestructive acoustic techniques together with artificial neural networks.

NDT assessment of bond quality between a protective coating and a concrete substrate is more difficult, because of the usually thin layer of new material involved and the differences in acoustic impedance. Garbacz and Garboczi (2003) have tried to evaluate bond between an epoxy coating and a C20/25 OPC concrete substrate with the UP-E method. They intended to determine whether an attenuation of the peak amplitude corresponding to a reflection from the interface could be correlated with bond strength. The value of attenuation was characterized using the amplification parameter $W_{0.8H}$ (amplification of the signal up to 0.8 of the full scale). The value of amplitude amplification needed to reach 0.8 of maximum amplitude (i.e. 80% of full screen height), called here as the coefficient, $W_{0.8H}$, was treated as a measure of epoxy coating adhesion to concrete substrate.

Analysis of Garbacz and Garboczi's results showed that there is no relation between the attenuation of the signal and the pull-off bond strength (Figure 7.6a). Furthermore, the value of $W_{0.8H}$ obtained for a delaminated epoxy coating was not significantly different from those recorded in well-bonded areas (bond strength value ranging between 0.5 and 3.0 MPa). The observed variations in the attenuation values result essentially from the coating thickness variations, as shown in Figure 7.6b.

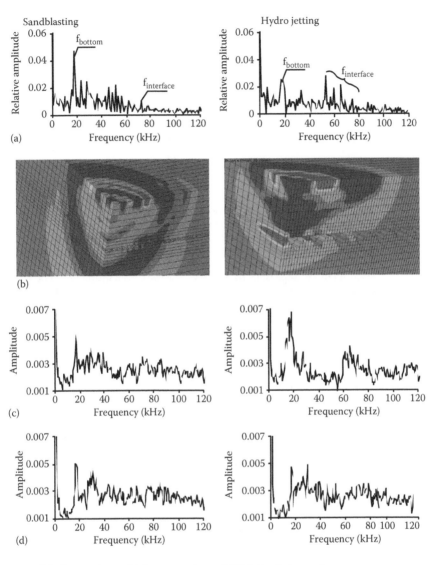

Figure 7.5 Comparison of I-E test results and FEM simulations of stress wave propagation across mortar repairs placed on test slabs following two types of surface preparation (sandblasting, hydro jetting): (a) experimental spectra for C40/50 concrete plate: 600 × 800 × 130 mm and repair mortar of 50 mm thick; results of FEM simulations of wave propagation in model repair—(b) examples of disturbance in wave propagation in the case of air voids present at the interface system and spectra obtained for concrete substrate with (c) unfilled irregularities, and (d) completely filled irregularities. (Adapted from Garbacz, A., *Non-Destructive Investigations of Polymer-Concrete Composites with Stress Waves—Repair Efficiency Evaluation*, Vol. 147, Publishing House of the Warsaw University of Technology, 2007, in Polish.)

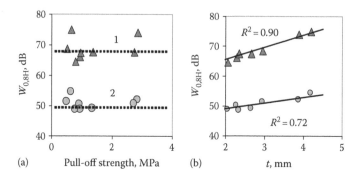

Figure 7.6 Survey of epoxy-coated concrete slabs with the UP-E method: amplification $W_{0.8H}$ versus (a) the pull-off strength and (b) epoxy coating thickness, with test results obtained in each case using two different commercial transducers SI2HB0.8-3 (●) and I0V202 (10 MHz frequency with delay line) (▲); dashed line—value of the amplification $W_{0.8H}$ for free (unbounded) 2-mm thick epoxy coating. (From Garbacz, A. and Garboczi, E.J., Ultrasonic evaluation methods applicable to polymer concrete composites, NISTIR 6975, National Institute of Standards and Technology, Gaithersburg, MD, 2003, www.nist.gov.org.)

Indirect ultrasonic pulse velocity (UPV) is another wave propagation test method, involving variable distance between transducers. While it is not commonly used, it has been applied for the detection of cracks in mortar repairs (Jaquerod et al., 1992), the evaluation of crack depth in reinforced concrete (e.g., Kaszyński, 2000), and the estimation of concrete degradation depth from the surface (Teodoru and Herf, 1996). Czarnecki et al. (2006) investigated its use for estimating the bond strength of various types of polymer coatings applied on concrete water-dispersed epoxy, solvent-dispersed epoxy, polyurethane, and vinylester-based. In this study, the methodological approach was based on the general assumption that weak or altered bond areas influence the recorded waveform. Two wave propagation parameters were used in the analysis, the pulse velocity and a the mean square value of amplitude in time domain ($MS(t)$). The MS value at a given point of waveform is calculated with the following formula:

$$MS(t) = \frac{\sum_{i=n_0}^{n} (A_i)^2}{(n - n_0)} \tag{7.2}$$

where
    A_i is the amplitude of the n_i-recorded point (V)
    n is the total number of datapoints recorded
    n_0 is the rank of the first data point with amplitude different from zero

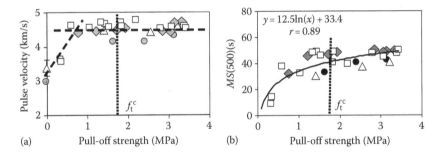

Figure 7.7 Evaluation of bond strength in epoxy-coated concrete substrate using the UPV method: pulse velocity (a) and mean square value of amplitude (MS) in time domain (500 μs) (b) vs. the pull-off strength for different polymer coatings: 2 mm thick water-dispersed epoxy (●), 2 mm thick solvent epoxy (△), 1 mm thick polyurethane (□), and 4 mm thick vinylester (◊); f_t^c —nominal tensile strength of the concrete substrate. (From Czarnecki, L. et al., *Cement Concr. Compos.*, 28(4), 360, 2006.)

The various concrete substrates tested by Czarnecki et al. (2006) were preconditioned at different moisture contents, and a bonding agent was applied on some of them in such a way to yield a wide range of bond characteristics. To simulate delamination (0-MPa bond strength), the polymer coating was prepared on a polyethylene sheet covered with an antiadhesive agent and then placed over the OPC concrete test substrate.

The test results presented in Figure 7.7 show that pulse velocity and *MS* correlate well with the recorded bond strength, both only up to a value of the order of 1 MPa. Above that theshold, both relationships are flattening. This might just reflect the nature of the pull-off test results, i.e., the maximum bond recordable strength corresponds to the tensile strength of the weakest of the two bonded materials (concrete susbtrate or repair material). Further investigation is necessary, but UPV could be used as a quantitative tool to determine whether the achieved bond strength has reached this limiting value.

7.2.2.5 Detection of bond discontinuities (zero-air volume defects)

The results presented in the previous subsection confirmed that *non-zero-volume* defects are relatively easy to detect with NDT methods such as I-E and UPV, provided that their size is significant. It is more complex to detect so-called *zero-volume* defects, e.g., dust, oil, or some other contaminants on a concrete substrate, which can alter the bond strength between the repair material and concrete as well.

A test program (Garbacz, 2015) intended to shed some light on the capability of detecting such bond defects with stress wave propagation methods was conducted on a series of OPC concrete (C20/25) test slabs that were preconditioned and uniformly sandblasted. Prior to overlaying with a proprietary polymer–cement mortar, discontinuities meant to simulate zero-volume interfacial defects were laid on two series of these test substrates, the third series being used as the reference. The artificial defects consisted respectively in the impregnation of a 350 × 550 mm area with demolding oil, and in a 200 × 400 mm plastic sheet. Each series of substrates was then subdivided into two-specimen subsets with respective overlay thicknesses of 50 and 80 mm.

All overlaid slabs underwent I-E soundings and interface peak amplitude mappings were generated. Defects having sufficient influence on bond continuity or integrity are expected to yield peaks corresponding to the interface depth in the frequency spectrum. Thus, the mappings should reveal areas where bond strength is lower. Figure 7.8 presents the amplitude distributions of the bottom and interface peaks recorded for the various repaired test slabs. It can be observed that the peak amplitude values recorded for the specimens enclosing an impregnated oiled area are close to those of the reference test slabs, indicating that such defect is difficult to

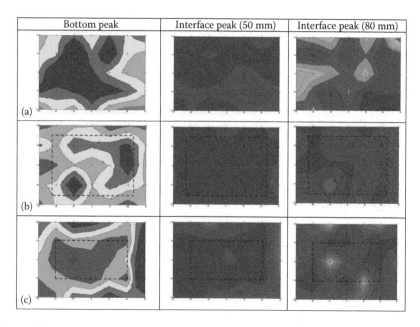

Figure 7.8 NDT survey of C40/50 overlaid concrete slabs (50 and 80 mm overlays), with or without artificially enclosed interfacial defects, using the I-E method- Amplitude distributions of bottom and interface peaks for the three investigated scenarios: (a) no defect (reference), (b) 350 × 550 mm area impregnated with demolding oil, and (c) 200 × 400 mm rectangular plastic sheet laid on the substrate. (From Garbacz, A., *Bull. Pol. Acad. Sci.: Tech. Sci.*, 63(1), 77, 2015.)

detect with impact echo. In the case of the repaired slab with a plastic sheet inserted at the interface, both the bottom and interface of peaks exhibited low amplitudes and low-frequency peaks, which are characteristic of shallow delamination.

The series of test slabs was investigated further using this time the ground-penetrating radar (GPR) technique, with 1.6 and 2.3 GHz antennas (Van der Wielen et al., 2010; Adamczewski et al., 2012). The results did not reveal a better ability of the GPR to detect such *zero-volume* interfacial defects in repair systems (Figure 7.9). Similar findings (Garbacz et al., 2013) were obtained in an investigation involving monolithic slabs containing

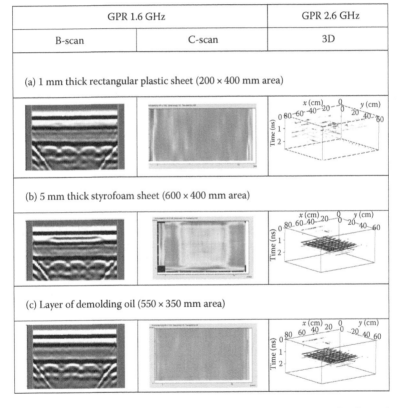

Figure 7.9 NDT survey of concrete overlaid slabs, with or without artificially enclosed interfacial defects, using the GPR with 1.6 and 2.6 GHz antennas—Test results for the three investigated scenarios: (a) 200 × 400 mm rectangular plastic sheet laid on the substrate, (b) 200 × 400 mm rectangular styrofoam plate of 5 mm thick, and (c) 350 × 550 mm area impregnated with demolding oil. (From Adamczewski, G. et al., Application of GPR method for the bond quality evaluation, *CD Proceedings, Proceedings of the 41st Polish Conference on Nondestructive Testing*, Toruń, Poland, 2012, in Polish.)

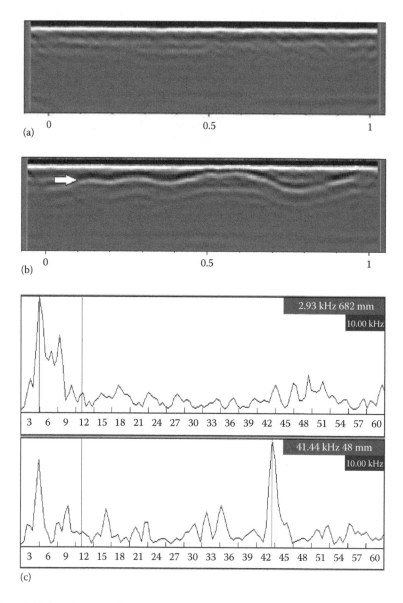

Figure 7.10 NDT survey of monolithic slabs containing artificial defects: (a) GPR B-scan image (cross-section view) of a slab with an enclosed polyethylene sheet, (b) B-scan image (cross-section view) of a slab with an enclosed air bubble sheet, and (c) typical I-E frequency spectra of shallow delamination (top) and interface peak amplitude (bottom) recorded for these slabs. (From Garbacz, A., *Adv. Mater. Res.*, 687, 359, 2013.)

either a smooth PVC foil or an air-bubble-filled PVC foil (Figure 7.10). In that study, the GPR only allowed to detect the air-bubble foil. In comparison, both foils were detected with the I-E method and most I-E signals (90%) revealed the presence of shallow delaminating.

7.3 INFLUENCE OF CONCRETE SURFACE CHARACTERISTICS

7.3.1 Mechanical integrity

As emphasized in ACI 546R-14 and EN 1504-10, the success of a concrete repair is highly dependent on adequate surface preparation. Ideally, the preparation of a concrete surface prior to repair should expose concrete with properties similar to those of the bulk concrete. The exposed concrete surface should be sound, uniform, cohesive, and free from dust, oil, or other contaminants, while and the shape of the surface should promote good anchorage for the repair material. Prior to repair, the existing concrete surface must thus be treated to remove deteriorated, contaminated and/or unsound material, as well as to roughen its profile.

While surface preparation is generally aimed at producing a sound substrate with properties and characteristics of the unaltered bulk concrete, it must be realized that the operation of concrete removal and preparation technique may itself induce defects within the substrate. Such an operation necessarily involves the transfer of a certain amount of energy to fracture the required quantity of material to be removed from the element and, depending on the technique, it will leave more or less significant mechanical defects (ranging from microcracking to macrocracking) in the residual concrete (Bissonnette et al., 2006). Superficial cracking, often referred to as "bruising," is considered as one of the most important parameters influencing adversely adhesion of concrete repairs and surface treatments. The prepared concrete surface needs to be free from significant microcracking; otherwise, the interfacial zone and the bond are inevitably altered. Hence, surface preparation can have detrimental co-lateral effects.

The amount of microcracking is governed by the method(s) selected for concrete removal. In general, mechanical methods involving repetitive concentrated impacting are likely to introduce microcracking (Silfwerbrand, 1990). A comparison between surfaces treated by hydro jetting and pneumatic hammers, respectively, is shown in Table 7.5. Because of microcracking, the bond strength was considerably lower in the case where the surface was treated with pneumatic hammers.

In a study intended to assess quantitatively the mechanical degradation induced during concrete removal and preparation operations and its effects upon bond strength, some of the most common surface preparation techniques were evaluated, namely, sandblasting, scarifying, high-pressure

Table 7.5 Laboratory pull-off tests: Influence of microcracking

Interface treatment	Presence of microcracking	All tests		Interface failures	
		Number of cores	Average failure stress (MPa)	Number of cores	Average failure stress (MPa)
Hydro jetting	No	16	1.86	1	2.23
Pneumatic hammers	Yes	16	1.10	5	0.94

Source: Silfwerbrand, J., *Concr. Int.*, 9, 61, 1990.

hydro jetting, and jackhammering (Bissonnette et al., 2006). The test specimen size (520 × 620 mm in plan and either 460 or 920 mm in depth) was chosen such as to simulate operations performed on massive elements and prevent the results to be affected by size effects. Depending on the surface preparation technique, the depth of concrete removed ranged from a few millimeters up to about 50 mm. The number of cracks and the total crack length resulting from the preparation with jackhammer were observed to be significantly larger than with any other of the investigated techniques. Moreover, it could be seen that increasing the jackhammer weight—and, thus, the impact energy—causes the extent of cracking (length and number of cracks) to increase substantially.

Pull-off tests were then performed on all the overlaid blocks in accordance with the ASTM C1583 standard procedure. The average results are summarized in Figure 7.11. Clearly, bond strength in direct tension can be considerably affected by the presence of induced microcracking in the substrate prior to repair: the bond strength values developed on surfaces prepared with jackhammers are much lower than on any of those prepared with the other investigated methods, and the heavier the hammer, the worse it gets. Overall, the cracking and bond strength results are quite consistent. As such, this tends to legitimate the concerns about surface preparation-induced microcracking and confirms the need for taking it into account in the selection of a concrete removal technique.

Still, some field tests show that bond strength can reach satisfactory values if removal with concrete breakers is followed by high-pressure water cleaning (Silfwerbrand and Petersson, 1993). Talbot et al. (1994) and Carter et al. (2002) concluded that sandblasting subsequent to the use of heavy impact methods could remove the damaged concrete and provide a satisfactory surface.

Besides, the use of suitable bonding agents can counteract the detrimental effect of bruising on the interface quality and bond strength by filling the microcracks and, to some degree, cementing the loosened concrete particles (Pareek et al., 1990; Garbacz et al., 2005; Bissonnette et al., 2006). However, in field conditions, it is not easy to guarantee the

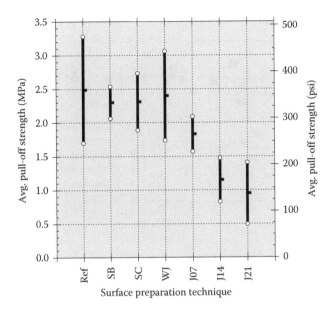

Figure 7.11 Average pull-off strength recorded for 75 mm (3 in.) ordinary concrete overlays cast on substrates treated with various surface preparation techniques (Ref: no preparation; SB, sandblasting; SC, scarifying; WJ, water jetting; J07, 7-kg jackhammering; J14, 14-kg jackhammering; J21, 21-kg jackhammering; repair concrete: CAN/CSA-Type 10 cement, 0.40 W/C, 10 mm coarse aggregate). (From Bissonnette, B., *Concr. Int.*, 28(12), 49, 2006.)

placement conditions required to achieve such a strengthening effect on a systematical basis.

7.3.2 Roughness

Mechanical interlocking is generally recognized as one of the governing mechanisms of adherence between existing concrete and a variety of surface treatments (Courard, 1998). Accordingly, surface roughness of the substrate has for long been considered to have a major influence on bond.

Many studies have been devoted to study the relationship between interface roughness and concrete repair bond strength.

In various reported investigations, such as those by Mainz and Zilch (1998) and Tschegg et al. (2000), the experimental data led to the conclusion that bond strength increases with the roughness of the substrate. Nonetheless, the existence of a threshold value was suggested by Takuwa et al. (2000), who studied the influence of surface preparation of the substrate concrete and defined a simple surface roughness parameter referred to as "increase of area" (Figure 7.12). Silfwerbrand (1990) compared interface strengths resulting from different surface treatments and a wide surface

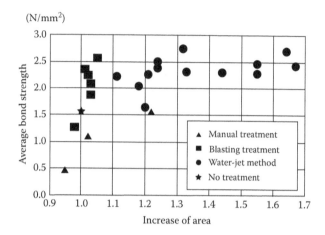

Figure 7.12 Relationship between surface roughness (described as increase of substrate area in percent) and bond strength. (From Takuwa, I. et al., *J. Jet Flow Eng.*, 17(1), 29, 2000.)

roughness range. He concluded that the threshold value for tensile bond strength improvement was in the range of the roughness values typical of sandblasted surfaces. An increase in surface roughness beyond this value would not increase any further bond strength. The existence of such a limiting roughness value was also suggested in other investigations (Garbacz et al., 2005; Czarnecki et al., 2007).

The actual influence of interface roughness on bond strength also depends on a range of other parameters such as material strength and effective surface area. Depending on the structure type and size, the nature of the work to be performed and the local construction and repair customs, a variety of surface treatments can be used and, as a consequence, a rather wide spectrum of surface roughness and characteristics can in fact be yielded. Therefore, interpretation of bond properties as a function of individual parameters should be considered with caution, as the relationships are non-univocal. For instance, an increase in roughness may be obtained at the expense of the substrate mechanical integrity, as suggested by the results displayed in Figure 7.13.

The cross influence of roughness and mechanical integrity of the substrate upon concrete repair bond has been studied by Bissonnette et al., (2014). The results of bond strength tests performed on repair test slabs that were prepared with a range of surface preparation techniques are presented in Figure 7.14. Except for the slabs prepared by jackhammering, it can be seen that the pull-off test results are close to the corresponding substrate splitting-tensile strength values for both the 20 and 35 MPa slab series. In the 20 MPa slabs, where it is particularly close, failure of the

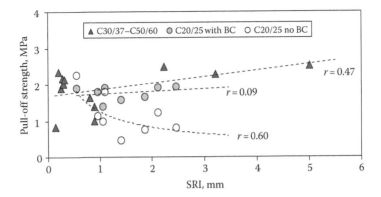

Figure 7.13 Pull-off strength versus SRI determined with sand patch test for a range of concrete substrates (concrete grades: C20/25 to C50/60) prepared with different surface preparation methods and repaired with a polymer-cement mortar; BC—bond coat (bonding agent). (From Czarnecki, L. et al., *Eng. Constr.*, 12, 630, 2007, in Polish.)

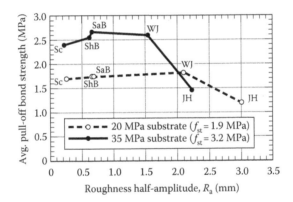

Figure 7.14 Results of pull-off tests (ASTM C1583) performed after repair of concrete test slabs treated with various surface preparation techniques (Sc, scarifying; ShB, shotblasting; SaB, sandblasting; HJ, hydro jetting; JH, jackhammering; repair material, 45 MPa OPC concrete). (From Bissonnette, B. et al., Concrete repair bond: Evaluation and factors of influence, *Proceedings of the Concrete Solutions: Fifth International Conference on Concrete Repair*, Belfast, Northern Ireland, 2014, pp. 51–57.)

pull-off specimens occurred systematically in the substrate (except again for the jackhammered slabs). On jackhammered slabs, even though lightweight hammers (7 kg) were used, the recorded pull-off strength values are significantly lower, and most of the time (>90%), failure occurred in the interface area. As for the corresponding weaker superficial tensile strength values (pull off experiments performed on the substrates, prior to repair), this has to be attributed to the presence of local defects left on the surface upon completion of the jackhammering operations.

As part of the same investigation carried out by Bissonnette et al. (2014), an artificially profiled test slab was cast to avoid the presence of preparation-induced damage and isolate the effect of roughness upon bond strength. V-shape rippled acrylic dies were installed at the bottom of the slab form to obtain triangle wave profiles with amplitude values of 2, 4, 6, and 8 mm respectively in four adjacent areas along the specimen length, the wavelength being of 30 mm in all of them. The direct tensile bond strength test results obtained for the artificially profiled slab (Figure 7.15) show that the average bond strength in tension is increasing with the substrate roughness amplitude. It clearly suggests that in absence of superficially induced damage, increasing the surface of contact leads to a stronger repair bond.

Hence, as far as the relationship between pull-off strength and substrate roughness is concerned, it appears that pull-off values slightly increase with the value of R_a, provided that no or limited damage is induced. Where the extent of damage induced by the surface preparation operations becomes significant, the positive influence of increased roughness can be completely offset by the adverse effects of bruising.

A quantitative analysis of the cross influence of surface roughness and macrocracking was carried out by Courard et al. (2014). Several repair systems have been tested on concrete slabs cast with different concrete grades and prepared with different surface preparation methods. The research program was divided in two test series. In the first test series (Group A),

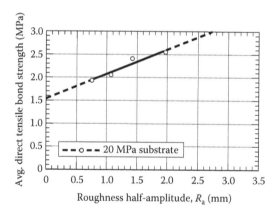

Figure 7.15 Results of direct tensile tests performed after repair on cores extracted from an artificially profiled 20 MPa concrete test slab. (From Bissonnette, B. et al., Concrete repair bond: Evaluation and factors of influence, *Proceedings of the Concrete Solutions: Fifth International Conference on Concrete Repair*, Belfast, Northern Ireland, 2014, pp. 51–57.)

three different OPC concrete mixtures (C30/37, C40/50, C45/55) and four types of surface preparation techniques were selected in order to obtain differences in surface profile and level of microcracking in the near-to-surface layer. The investigated surface preparation were polishing (PL), dry sandblasting (SB-D), jackhammering (JH), and high-pressure (250 MPa) hydro jetting (HJ). In the second test series (Group B), three more concrete mixtures (C25/30, C35/45, C50/60) and less aggressive surface preparation techniques were selected such as to obtain similar profiles and limited microcracking. The following surface preparation techniques were chosen, in view of generating profiles similar to those tested in the first series, but with less microcracking: brushing (BR), wet sandblasting (SB-W), scarifying (SC), and low-pressure (12 MPa) hydro jetting (LC). Surface roughness was characterized in accordance with EN 1766, with the volumetric sand patch technique (see Chapter 3). In selected specimens, cracking was assessed through microscope examinations. Based on the examination records, the density of microcracks (L_A) was calculated according to the following equation:

$$L_A = \frac{l_A}{A} \ (\text{mm/cm}^2) \tag{7.3}$$

where

$l_A = \Sigma l$—total microcrack length (mm) measured on photographs
$A = d_0 \times l_0$—observation area: 2.0 cm × 8.0 cm = 1.6 cm² (d_0, thickness of the observation area beneath the profile line; l_0, projection length of the profile)

In addition, the superficial concrete integrity after surface preparation was characterized in accordance with EN 1504-10, by evaluating the surface tensile strength (f_{hs}) through pull-off testing (EN 1542). After extensive surface characterization of the prepared slabs, they were overlaid with different commercial polymer–cement repair mortars and submitted to pull off tests after 28 days of curing. To analyze the relationships between the surface parameters (surface roughness index [SRI], L_A, f_{hs}) and bond strength (f_h), a multiple regression approach was applied. Groups A and B are not directly compared as test slabs were prepared with different concrete mixtures and surface preparation methods. The data treatment revealed that the most influential parameters in the prediction of bond strength (f_h) are f_{hs} and SRI. The corresponding results of the multiple regression approach for both test series are presented in Figure 7.16. Again, the trends observed show overall that repair bond strength increases with the level of roughness of the substrate, provided that the mechanical integrity in the latter is not compromised by surface preparation defects. The opposite trends found between Groups A and B for the relationship between bond strength and the superficial tensile strength of the substrate reveal the

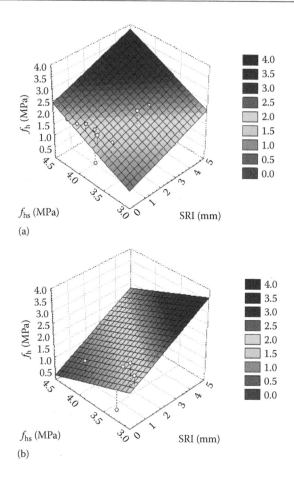

Figure 7.16 Bond strength test results (fh) as a function of the substrate superficial tensile strength (fhs) and roughness (SRI) for test series (a) Group A and (b) Group B, where test slabs prepared with different concrete grades and submitted to a range of surface preparation methods were overlaid with commercial polymer-cement repair mortars. (From Courard, L. et al., *Cement Concr. Compos.*, 46, 73, 2014.)

complex interaction between the numerous characteristics and phenomena involved, though it does not invalidate the conclusions regarding the cross influence of roughness and mechanical integrity of the substrate.

Beushausen (2005) stated that interface roughness exerts a significant influence on shear bond strength, but it only plays a minor role in the tensile bond behavior. According to Rilem TC-RLS (2011), further investigations using interface shear bond tests are recommended to validate the substantiate the assumption of different bond mechanisms. In a comprehensive study, Bissonnette et al. (2014) also performed torsional shear bond tests.

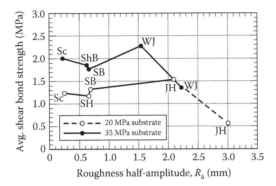

Figure 7.17 Results of torsional bond experiments performed after repair of concrete test slabs treated with various surface preparation techniques (Sc, scarifying; ShB, shotblasting; SB, sandblasting; HJ, hydro jetting; JH, jackhammering; repair material, 45 MPa OPC concrete). (From Bissonnette, B. et al., Concrete repair bond: Evaluation and factors of influence, *Proceedings of the Concrete Solutions: Fifth International Conference on Concrete Repair*, Belfast, Northern Ireland, 2014, pp. 51–57.)

The results presented in Figure 7.17 actually show much similarity with the tensile pull-off data plotted in Figure 7.14, both in terms of magnitude and trends. Again the substrate strength and the presence of damage are influential parameters. Still, roughness does not appear to play a more important role in shear than in tension.

7.3.3 Saturation level

The substrate moisture condition may exert a significant influence on bond characteristics of surface treatments applied to concrete. For some systems, the presence of moisture is desirable, while for some others, such as some epoxy-based materials, the surface needs to be dry at the time of placement.

The influence of surface moisture on bond between existing concrete and repair is an issue of significant importance and has been the subject of a number of studies and much debate over the years (Rilem TC-RLS, 2011). On the one hand, dry, "thirsty" concrete surface tends to suck water from the overlay, which may result in a weak interfacial repair layer and lower bond strength. On the other hand, a surface that is too wet tends to dilute the repair material at the interface and increases the water/cement ratio and hence leads to low material strength, increased shrinkage, and reduced bond strength. Water in open pores further prevents the interlocking effect, while free water at the surface substrate may destroy the bond completely.

In general, the opinions on the effects of substrate moisture differ significantly among the scientific community (Pigeon and Saucier, 1992). For instance, Zhu (1992) found experimental signs of optimal moisture, but the moisture influence on the bond was so small that it was difficult to discern between moisture influence and scatter of test results. Conversely, Li et al. (1999) measured the bond strength of repaired specimens after freeze–thaw cycles and found that the optimum interface moisture conditions at the time of casting depends on the nature and characteristics of the repair material.

Being considered by many as the best compromise (Saucier and Pigeon, 1991), *saturated surface dry* (SSD) conditioning of the substrate prior to application of cementitious repair materials is typically recommended and specified, which underlies the "layman's" instinctive procedures to avoid problems, rather than achieving the most effective bond. Various investigators came to the conclusion that different substrates and repair materials may require different interface moisture conditions at the time of casting to achieve optimum interfacial bond. The problem is that presently there is no test method to determine the optimum moisture condition for a given combination of substrate and repair material.

Water is one of the critical factors influencing bond development between concrete and repair materials: it may accumulate at the interface or migrate through it in either direction, as a result of mechanical (i.e., gravity), chemical (i.e., hydration), or physical (i.e., temperature gradients) driving forces (Courard, 2000). Moisture condition of the concrete substrate surface at the time of repair material directly affects the moisture transport mechanisms between the freshly applied material and the existing concrete substrate. In fact, different moisture transport parameters affect the formation and behavior of the repair interfacial zone, such as diffusion and permeability coefficients, the interface characteristics being indeed influenced by different forms of water interaction:

- First, moist conditioning of the substrate before the application of the repair system is a key consideration; partial or total saturation of a concrete substrate is a common situation in repair works; water along the interface may prevent adhesion to the repair system, depending on the nature and characteristics of the repair material (Courard et al., 1999).
- Second, water or aqueous solution movements may appear (Courard et al., 2003) due to migration and infiltration along the interface (Courard and Darimont, 1998) or diffusion and capillary absorption from the zones to be repaired (Courard et al., 1999); the importance of these water movements will directly depend on the quality of the materials (i.e., composition, porosity).

In most situations, the saturation level at the interface appears to be a predominant factor in achieving strong and lasting adhesion of the repair system.

Recent findings from a comprehensive study (Courard et al., 2011) evidenced the effect of water in the substrate concrete superficial zone and

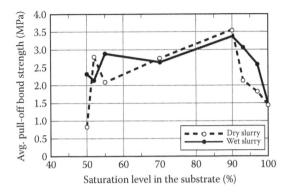

Figure 7.18 Pull-off test results recorded for a polymer-modified repair mortar cast over concrete substrates at various saturation levels, with dry or wet polymer-modified slurry. (From Courard, L., *Constr. Build. Mater.*, 25(5), 2488, 2011.)

the difficulty encountered in evaluating reliably the actual saturation level. The influence of the substrate moisture content upon bond strength is illustrated for a polymer-modified repair mortar in Figure 7.18. Overall, for the repair systems considered in the test program, it appears that optimum saturation levels for repair bond strength would lie somewhere between 70% and 90%.

Clearly, additional work is required to identify a methodology that could be used in field applications and, furthermore, to assess more precisely and reliably what are the optimum moisture ranges for different types of repair materials.

7.3.4 Cleanliness

Surface cleanliness is probably the most crucial factor in concrete repairs, as any loose debris, dirt, grease, and other surface contaminants can act as bond breakers (Bissonnette et al., 2012).

A surface that is contaminated at the time of overlay placement will produce poor bond characteristics (Rilem TC-RLS, 2011). In the first Swedish bridges repaired with high-pressure hydro jetting and bonded overlays in 1984 and 1985, coring showed poor bond at several locations. In most cases, insufficient cleaning was the reason for this (Silfwerbrand, 1990). Loose particles were found in the interface between old and new concrete.

In order to warrant cleanliness, the prepared surface should be cleaned twice (Rilem TC-RLS, 2011). The first cleaning step should be performed shortly after hydro jetting in order to prevent loose concrete particles, such as exposed unhydrated cement surfaces, to bond to the surface. The second cleaning step should be carried out prior to overlay placement to make sure that the surface is free from sand, oil, dust, or other particles originating from the environment or the construction operations. Hosing down with high-pressure water and the vacuum cleaning have proven to be most effective methods.

These considerations on cleanliness highlights the importance of workmanship (Bissonnette et al., 2012), as all the means taken to ensure bond quality and durability can easily be cancelled out by improper field practice.

7.3.5 Surface composition heterogeneity

In many cases of adhesive bond failure, debonding develops preferentially at the interfaces between the repair material and the exposed aggregates in the substrate. SEM observations actually reveal different local interfacial characteristics depending on the area (local composition) where the new material is in contact with the heterogeneous concrete substrate. (Figure 7.19).

(a)

(b)

Aggregate in the substrate covered with a thin film of $Ca(OH)_2$

Repair material in contact with the substrate mortar

(c)

Figure 7.19 Fracture surface located in the interfacial area of a concrete substrate overlaid after grinding (a) adhesive bond failure, (b) observation of the fracture surface under horizontal light, and (c) SEM micrograph. (From Courard, L., *Mag. Concr. Res.*, 57(5), 273, 2005.)

If the mechanical performance of a bonded composite system involving a heterogeneous material simply depended on the cumulative contributions of each of its individual paired components, with respect to their relative volume or surface, the resulting adhesion between a repair material (or resin) and concrete substrate would be equal to the weighed summation of the repair material paste (or resin) to concrete paste adhesion and repair material paste (or resin) to aggregate adhesions (respective contributions of the order of 35 % and 65 % for ordinary concrete). Nevertheless, it has been found experimentally that this simplistic *law of mixture* approach is not valid for concrete, the strength of the composite being in general significantly larger (of the order of 40 %) than the cumulative weighed contributions of each paired component (Courard, 2005).

In-depth characterization of the concrete surface composition actually evidenced that the *interfacial transition zone* (ITZ) at aggregate periphery, which can be up to 10 times more porous than cement paste, has to be taken into account in evaluating the surface porosity of the substrate available for the physicochemical anchorage of the new layer (Tabor, 1981; Larbi and Bijen, 1991). This hypothesis is supported further by microscopical observations, which revealed deeper methylene blue dye penetration from the concrete surface around the aggregates (Courard, 1998).

Taking into account the number and the dimensions of the aggregates at the surface, it is possible to evaluate the cumulative ITZ volume (Courard, 2005). Although it is likely overestimated, due to ITZ overlapping between adjacent grains, the calculation emphasizes the importance of this particular surface area in the adhesion of repair systems (as well as that of a variety of surface treatment systems for concrete). Anchorage of the repair material cement paste in the ITZ's, between cement paste and aggregates (Figure 3.37), increases the effective surface of contact between the concrete substrate and the repair material, leading to higher adhesion than what could be expected from the simple addition of the paste-paste bond and paste-aggregate bond.

This phenomenon is particularly evident when observing a fracture surface located within the repair interfacial area (Fig. 7.19 a,b): the layer of new material adhering to the substrate concrete cement paste is most generally thicker than that found on the aggregates. The deeper penetration of the new material inside the porosity of the superficial layer of concrete substrate at the periphery of the aggregates contributes significantly to the development and strength of interfacial bond.

7.3.6 Contamination

The effectiveness of mechanical adhesion of the repair material to the prepared concrete substrate relies on the ability of the cement paste to fill the surface profile valleys and penetrate in the open pores, such as to develop cohesion interlocking and anchoring effects.

Carbonation occurs when constituents in the concrete react with carbon dioxide and moisture to produce calcium carbonate. It typically advances inward from the exposed concrete surface. The outer layer of concrete in which an appreciable reaction with carbon dioxide has taken place is called the carbonated layer. Thus, the phenomenon of carbonation, which produces a denser surface layer with a so-called *clogged* pore system, reduces the absorptivity of the substrate concrete and might be expected to negatively affect bond strength.

Carbonation is generally a very slow process in good quality concrete, typically of the order of 1 mm/year. For typical ambient temperature ranges, the rate of carbonation is greater at relative humidity values ranging from 50% to 75%, although the relative humidity at which the maximum rate of carbonation is observed may be greater in concretes with higher porosity.

Cracks, microcracks, or any other defects in the concrete allow carbon dioxide easier access through the concrete surface, and carbonation can occur. The active coefficient of carbon dioxide diffusion in a concrete crack, 0.008 in. (0.2 mm) wide, is about three orders of magnitude (1000 times) higher than in average quality, crack-free concrete (Alekseev and Rosenthal, 1976).

Even though proper concrete removal and surface preparation operations usually eliminate carbonated concrete, relatively long periods of time between surface preparation and repair material placement may result in carbonation of the exposed substrate surface. The results of a study conducted at the U.S. Bureau of Reclamation (Denver, CO) indicated that freshly sandblasted concrete surfaces can show signs of carbonation within a few hours (Bissonnette et al., 2012).

According to Schrader (1992a), carbonation of the substrate can result in a soft surface and dusting, which may result in poor bond strength if an overlay is applied. Similar test results were obtained by Gulyas et al. (1995), who found that substrate carbonation can decrease bond significantly. By contrast, Block and Porth (1989) found that substrate carbonation does not affect pull-off bond strength. These contradicting results show the problems inherent in interpreting bond test results for complex systems in terms of a single test parameter. The actual differences in results can be explained by likely differences in surface preparation, repair material application, and curing.

A test program intended to shed new light on the effect of carbonation upon bond development in repair systems was conducted at the U.S. Bureau of Reclamation recently (Bissonnette et al., 2014). After being prepared respectively by sandblasting and by chipping hammer, two slab test series underwent carbonation in strictly controlled conditions until a depth of contamination of 3 mm was reached. In each series, slabs were protected from carbonation with properly sealed plastic sheet to serve as control. All test slabs were then overlaid with a 100 mm thick layer of 28 MPa concrete and submitted to bond testing after proper curing. The test results are displayed in Figure 7.20.

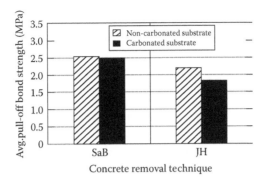

Figure 7.20 Pull-off test results recorded for test slabs repaired with 28 MPa concrete after different types of surface preparation, with or without carbonation (SaB, sandblasting; JH, jackhammering; repair material, 28 MPa OPC concrete). (From Bissonnette, B. et al., Concrete repair bond: Evaluation and factors of influence, *Proceedings of the Concrete Solutions: Fifth International Conference on Concrete Repair*, Belfast, Northern Ireland, 2014, pp. 51–57.)

For substrate surfaces prepared by sandblasting, no difference in bond strength is observed between carbonated and noncarbonated concrete substrates. Conversely, for substrates prepared with a concrete breaker, a significant reduction (16%) in bond strength is found for carbonated surfaces as compared to noncarbonated surfaces. Such different effects of carbonation could be attributed to the possible microdefects (bruising) of the surface prepared by jackhammering. Testing performed using only one type of a repair material does not allow to draw general conclusions about the overall effect of carbonation on tensile bond strength. Different repair materials may not necessarily behave the same way in terms of bond development to the carbonated surfaces. It appears likely, though, that carbonation may have only a slight impact on bond strength for an otherwise sound, properly prepared concrete substrate surface.

7.3.7 Bonding agents

The use of bonding agents has been a relatively common practice intended to ensure adequate bond of overlays and concrete repairs (Silfwerbrand, 1992; Bissonnette et al., 2013). The variety of materials available for such applications includes cement grout and mortar, latex and other polymer materials, latex-modified grout and mortar, and epoxy resins. An important aspect to consider when selecting a bonding agent is its compatibility with the materials to bond. Although no single material used as a bonding agent has proved to consistently yield durable bond, the most conclusive results are still associated with the use of cement grout. Ideally, the grout should be sprayed onto the substrate in order to maximize the uniformity of application. Spraying should be performed on an SSD surface, at a pace such that a distance of 2.5–3.0 m is kept in front of the overlay material placement. The fresh bonding agent layer should be spread uniformly in such a way to prevent accumulations in low points of the surface. Drying of the slurry before it is overlaid must be prevented, as it can affect bond significantly. If its color becomes lighter, a slight amount should be added before overlaying. In areas where it has started hardening, it should be removed (through scabbling or sandblasting) and reapplied. The main advantage of using a bonding agent does not lie in a significant increase of the average bond strength but mostly in a more uniform and reliable bond over the surface. This is true only if the agent is placed adequately and in a timely manner, which often proves to be difficult in the field.

Garbacz et al. (2006) analyzed the bond strength development of specific repair systems placed on substrates with different surface roughness characteristics, with and without a bonding agent (Garbacz et al. 2006). The relationships between the pull-off strength and selected surface roughness parameters were not found to be statistically significant for both overlay systems investigated, i.e., a polymer-based repair mortar used with or

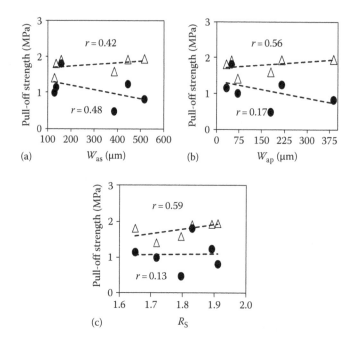

Figure 7.21 Relationships between the pull-off strength and the mean value of waviness profiles determined with (a) laser (Was), (b) mechanical profilometry (Wap), and (c) the surface roughness ratio R_S; repair systems with ($\triangle$) and without ($\bullet$) bonding agent; C20/25 OPC concrete. (From Garbacz, A. et al., *Mater. Charact.*, 56, 281, 2006.)

without a bonding agent (Figure 7.21). Still, it could be observed that for repairs placed with a bonding agent, the pull-off strength slightly increases as the surface roughness increases, while the opposite was observed for the repairs without bonding agent. This can be explained by the fact that the investigated repair mortar had relatively low workability (in part due to its fiber content) and could not wet adequately the substrate (Figure 7.22). Characterized by much better rheological characteristics, the bonding agent could much more effectively fill the profile valleys and penetrate into the surface porosity. Bonding agents can thus improve bond strength for repair materials having inadequate rheological characteristics. The ability of the repair system to adequately wet the irregular surface of the substrate is a very important factor with regard to bond development. Besides, in some circumstances, the bonding agent may cement loose or bruised particles remaining on an inadequately cleaned surface and prevent or limit their adverse effect on bond development.

Nevertheless, bonding agents cannot compensate for improper substrate surface preparation and may act as a bond breaker when used inappropriately (Pigeon and Saucier, 1992; Schrader, 1992b). Although they are still regularly specified, the body of data published over the years never led to conclusive evidence of their actual usefulness in the case of cement-based repair concretes and mortars. This is most likely related to the increased risks of compatibility problems incurred when introducing a third material and a second interface in the already intricate repair composite systems.

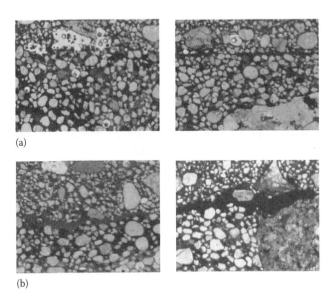

(a)

(b)

Figure 7.22 View of the interface between a polymer-based repair material and concrete substrate after (a) sandblasting and (b) milling without (left) and with (right) bonding agent. (Adapted from Garbacz, A. et al., *Mag. Concr. Res.*, 57, 49, 2005.)

As a matter of fact, with the surface preparation and placement techniques and materials available today, the use of bonding agents is generally not required anymore to generate proper bond for cement-base repair materials (ACPA, 1990). The importance of the quality of surface preparation operations was addressed previously. Also, the availability of vibrating placement equipment today contributes to the achievement of satisfactory bond on a more consistent basis, as vibration minimizes voids in the interface area and promotes penetration of the repair material into the substrate pores.

7.4 CONCLUSION

There are many purposes for repair and surface treatment of concrete structures and elements, such as prolonging the useful service life, preserving the architectural qualities, restoring the load-carrying capacity, or strengthening. For the work to be effective and meet the objectives, lasting bond between the newly applied material and the concrete substrate is essential. A critical requirement in achieving durable bond is adequate surface preparation and conditioning of the substrate prior to application of repair material or surface treatment. Regardless of the quality of the material and placement operations, the quality of surface preparation of the concrete substrate will often determine whether the intervention is a success or a failure.

REFERENCES

ACI/ICRI (2013) Concrete Repair Manual, American Concrete Institute, Farmington Hills, MI, 862p.

ACPA (1990) Guidelines for bonded concrete overlays. Technical Bulletin TB-007P, American Concrete Pavement Association, Arlington Heights, IL, 15p.

Adamczewski, G., Garbacz, A., Van der Wielen, A., Piotrowski, T., and Courard, L. (2012) Application of GPR method for the bond quality evaluation. *CD Proceedings, Proceedings of the 41st Polish Conference on Nondestructive Testing*, Toruń, Poland (in Polish).

Adams, R.D. and Drinkwater, B.W. (1997) Nondestructive testing of adhesively bonded joints. *NDT&E International*, 30, 93–98.

Alekseev, S.N. and Rosenthal, N.K. (1976) *Resistance of Reinforced Concrete in Industrial Environment*. Moscow, Russia: Stroyisdat.

ACI 546 R-14 (2014). Guide to concrete repair. Farmington Hills, MI: American Concrete Institute, 70pp.

ASTM C1042-99 (1999). Standard test method for bond strength of latex systems used with concrete by slant shear (withdrawn 2008). West Conshohocken, PA: ASTM International, 3pp.

ASTM C1404/C1404M-98 (2003). Standard test method for bond strength of adhesive systems used with concrete as measured by direct tension (withdrawn 2010). West Conshohocken, PA: ASTM International, 5pp.

ASTM D4541-09e1 (2009). Standard test method for pull-off strength of coatings using portable adhesion testers. West Conshohocken, PA: ASTM International, 16pp.

ASTM C794-15 (2015). Standard test method for adhesion-in-peel of elastomeric joint sealants. West Conshohocken, PA: ASTM International, 7pp.

ASTM D903-98 (2010). Standard test method for peel or stripping strength of adhesive bonds. West Conshohocken, PA: ASTM International, 3pp.

ASTM C1583/C1583M-13 (2013). Standard test method for tensile strength of concrete surfaces and the bond strength or tensile strength of concrete repair and overlay materials by direct tension (pull-off method). West Conshohocken, PA: ASTM International, 5pp.

ASTM C882/C882M-13a (2013). Standard test method for bond strength of epoxy-resin systems used with concrete by slant shear. West Conshohocken, PA: ASTM International, 4pp.

ASTM C957/C957M-15 (2015). Standard specification for high-solids content, cold liquid-applied elastomeric waterproofing membrane with integral wearing surface. West Conshohocken, PA: ASTM International, 4pp.

Beushausen, H.D. (2005) Long-term performance of bonded overlays subjected to differential shrinkage. PhD Thesis, University of Cape Town, Cape Town, South Africa, 264pp.

Bissonnette, B., Courard, L., Beushausen, H., Fowler, D.W., and Vaysburd, A.M. (2013) Recommendations for the repair, the lining or the strengthening of concrete slabs or pavements with bonded cement-based material overlays. *Materials and Structures*, 46, 481–494.

Bissonnette, B., Courard, L., Fowler, D.W., and Granju, J.L. (2011) Bonded cement-based material overlays for the repair, the lining or the strengthening of slabs and pavements. RILEM STAR Report Volume 3, 193-RLS RILEM TC, Springer, Dordrecht, the Netherlands, 175p.

Bissonnette, B., Courard, L., Garbacz, A., Vaysburd, A., and von Fay, K. (2014) Concrete repair bond: Evaluation and factors of influence. *Proceedings of the Concrete Solutions: Fifth International Conference on Concrete Repair*, Belfast, Northern Ireland, pp. 51–57.

Bissonnette, B., Courard, L., Vaysburd, A.M., and Bélair, N. (2006) Concrete removal techniques—Influence on residual cracking and bond strength. *Concrete International*, 28(12), 49–55.

Bissonnette, B., Vaysburd, A.M., and von Fay, K.F. (2012) Best practices for preparing concrete surfaces prior to repairs and overlays. Report Number MERL 12-17, U.S. Bureau of Reclamation, Denver, CO, 92p.

Block, K. and Porth, M. (1989) Spritzbeton auf Carbonatisertem Beton. *Beton*, 7, 299–302.

CAN/CSA-A23.2-6B-14 (2014). Determination of bond strength of bonded toppings and overlays and of direct tensile strength of concrete, mortar, and grout. Rexdale, ON, Canada: Canadian Standards Association, 6pp.

Carter, P., Gurjar, S., and Wong, J. (2002) Debonding of highway bridge deck overlays. *Concrete International*, 24(7), 51–58.

Courard, L. (1998) Contribution à l'analyse des paramètres influençant la création de l'interface entre un béton et un système de réparation. Appétence et adhérence: cause et effet d'une liaison. PhD Thesis, University of Liège, Liège, Belgium, 192pp.

Courard, L. (2000) Parametric study for the creation of the interface between concrete and repair products. *Materials and Structures*, **33**(1), 65–72.

Courard, L. (2005) Adhesion of repair systems to concrete: Influence of interfacial topography and transport phenomena. *Magazine of Concrete Research*, **57**(5), 273–282.

Courard, L. and Darimont, A. (1998) Appetency and adhesion: Analysis of the kinetics of contact between concrete and repairing mortars. In: *Proceedings of the RILEM International Conference, Interfacial Transition Zone in Cementitious Composites*, Haïfa, Israel (Eds. A. Katz, A. Bentur, M. Alexander, and G. Arliguie), E&FN Spon, pp. 185–194.

Courard, L., Degeimbre, R., Darimont, A., and Wiertz, J. (1999) Effects of sunshine/rain cycles on the behaviour of repair systems. In: *Proceedings of the Second International RILEM Symposium on Adhesion between Polymers and Concrete, ISAP'99*, Dresden, Germany (Eds. Y. Ohama and M. Puterman), pp. 511–521.

Courard, L., Degeimbre, R., Darimont, A., and Wiertz, J. (2003) Hygrothermal application conditions and adhesion. In: *Proceedings of the Fifth International Colloquium Industrial Floors* (Ed. P. Seidler), Technische Akademie Esslingen, Ostfildern/Stuttgart, Germany, pp. 137–142.

Courard, L., Lenaers, J.F., Michel, F., and Garbacz, A. (2011) Saturation level of the superficial zone of concrete and adhesion of repair systems. *Construction and Building Materials*, **25**(5), 2488–2494.

Courard, L., Piotrowski, T., and Garbacz, A. (2014) Near-to-surface properties affecting bond strength in concrete repair. *Cement and Concrete Composites*, **46**, 73–80.

Czarnecki, L., Courard, L., and Garbacz, A. (2007) Application of surface engineering methods towards evaluation of concrete repair efficiency. *Engineering and Construction*, **12**, 630–634 (in Polish).

Czarnecki, L., Garbacz, A., and Krystosiak, M. (2006) On the ultrasonic assessment of adhesion between polymer coating and concrete substrate. *Cement and Concrete Composites*, **28**(4), 360–369.

EN 1504-10:2005 (2005). Site application of products and quality control of the works, Products and systems for the protection and repair of concrete structures—definitions, requirements, quality control and evaluation of conformity. Brussels, Belgium: European Standardization.

EN 1542:1999 (1999). Products and systems for the protection and repair of concrete structures. Test methods. Measurement of bond strength by pull-off Brussels, Belgium: European Standardization.

Garbacz, A. (2007) *Non-Destructive Investigations of Polymer-Concrete Composites with Stress Waves—Repair Efficiency Evaluation*, Vol. 147. Publishing House of the Warsaw University of Technology, Warsaw, Poland (in Polish).

Garbacz, A. (2010) Stress wave propagation throughout an interface: PCC composites-concrete substrate in repair system. *ACEE*, **3**(3), 35–44.

Garbacz, A. (2015) Application of stress based NDT methods for concrete repair bond quality control. *Bulletin of the Polish Academy of Sciences: Technical Sciences*, **63**(1), 77–86.

Garbacz, A., Courard, L., and Kostana, K. (2006) Characterization of concrete surface roughness and its relation to adhesion in repair systems. *Materials Characterization*, **56**, 281–289.

Garbacz, A. and Garboczi, E.J. (2003) Ultrasonic evaluation methods applicable to polymer concrete composites. NISTIR 6975, National Institute of Standards and Technology, Gaithersburg, MD.

Garbacz, A., Górka, M., and Courard, L. (2005) On the effect of concrete surface treatment on adhesion in repair systems. *Magazine of Concrete Research*, 57, 49–60.

Garbacz, A. and Kwaśniewski, L. (2006) Modeling of stress wave propagation in repair systems tested with impact-echo method. *Proceedings of the Brittle Matrix Composites*, Vol. 8, Warszawa, Poland, pp. 303–314.

Garbacz, A., Piotrowski, T., Adamczewski, G., and Załęgowski, K. (2013) UIR-scanner potential to defect detection in concrete. *Advanced Materials Research*, 687, 359–365.

Gulyas, R.J., Wirthlin, G.J., and Champa, J.T. (1995) Evaluation of keyway grout test methods for precast concrete bridges. *PCI Journal*, 40(1), 44–57.

Hoła, J., Bień, J., Sadowski, Ł., and Schabowicz, K. (2015) Non-destructive and minor-destructive diagnostics of concrete structures in assessment of their durability. *Bulletin of the Polish Academy of Sciences: Technical Sciences*, 63(1), 87–96.

ICRI No. 210.3-2013 (2013). Guide for using in-situ tensile pulloff tests to evaluate bond of concrete surface materials. Rosemont, IL: International Concrete Repair Institute, 20pp.

Jaquerod, C., Chippis, Ch., Alou, F., and Houst, Y.F. (1992) Nondestructive testing of repair mortars for concrete. *Proceedings of the Third International Colloquium Materials Science and Restoration*, Esslingen, Germany, pp. 872–888.

Kaszyński, J. (2000) An investigation into the depth in cracks in a reinforced concrete structures using an ultrasonic method. *Proceedings of the 15th World Conference on Non Destructive Testing*, Roma, Italy, p. 357.

Kwaśniewski, L. and Garbacz, A. (2008) Characterization of stress wave propagation in impact-echo method using FEM models of repair systems. *Proceedings of the International Conference on Challenges for Civil Construction*, FEUP, Porto, Portugal, pp. 92–93 + CD.

Larbi, J. and Bijen, J.M. (1991) The role of the cement paste-aggregate interfacial zone on water absorption and diffusion of ions and gases in concrete. In: *The Cement Paste Aggregate Interfacial Zone in Concrete* (Ed. J.M. Bijen), Technische Universiteit, Delft, the Netherlands, pp. 76–93.

Lewińska-Romicka, A. (2000) *Nondestructive Testing. Fundamentals of Defectoscopy*. Warszawa, Poland: WNT (in Polish).

Li, S.E., Geissert, D.G., Frantz, G.C., and Stephens, J.E. (1999) Freeze-thaw bond durability of rapid-setting concrete repair materials. *ACI Material Journal*, 96(2), 241–249.

Mainz, J. and Zilch, K. (February 1998) Schubtragfähigkeit von Betonergänzungen an nachträglich aufgerauhten Betonoberflächen bei Sanierungs-und Ertüchtigungsmassnahmen. Research report, Technical University Munich, Munich, Germany.

Malhotra, V.M. and Carino, N.J. (2004) *Handbook on Nondestructive Testing of Concrete*. CRC Press.

Maso, J.-C. (1980) Bonding between aggregates and hydrated cement paste. *VIIth International Congress on Chemistry of Cements*, Septima, edn., Vol. 18, Paris, France, pp. 61–64 (in French).

Michigan Department of Transportation (MDOT). Direct shear bonding test-qualification procedure for prepackaged hydraulic patching mortars, 1p. (after American Concrete Institute (June 2014)).

Pareek, S.N., Ohama, Y., and Demura, K. (1990) Adhesion mechanism of ordinary cement mortar to mortar substrates by polymer dispersion coatings. *Proceedings of the Sixth ICPIC'1990*, Shanghai, China, pp. 442–449.

Pigeon, M. and Saucier, F. (1992) Durability of repaired concrete structures. *Proceedings of the International Symposium on Advances in Concrete Technology*, October 11–12, Athens, Greece, pp. 741–773.

Sadowski, Ł. and Hoła, J. (2014) New nondestructive way of identifying the values of pull-off adhesion between concrete layers in floors. *Journal of Civil Engineering and Management*, 20(4), 561–569.

Sansalone, M. and Lin, J.M. (1994a) Impact-echo response of hollow cylindrical concrete structures surrounded by soil and rock. Part I—Numerical studies. *ASTM Geotechnical Testing Journal*, 17, 207–219.

Sansalone, M. and Lin, J.M. (1994b) Impact-echo response of hollow cylindrical concrete structures surrounded by soil and rock. Part II—Field studies. *ASTM Geotechnical Testing Journal*, 17, 220–226.

Santos, P., Júlio, E., and Santos, J. (2011) Towards the development of an in situ non-destructive method to control the quality of concrete-to-concrete interfaces. *Engineering Structures*, 32(1), 207–217.

Saucier, F. and Pigeon, M. (1991) Durability of new-to old concrete bondings. *Proceedings of the ACI International Conference on Evaluation and Rehabilitation of Concrete Structures and Innovations in Design*, ACI SP-128, December, Hong Kong, China, pp. 689–705.

Schrader, E.K. (1992a) Mistakes, misconceptions, and controversial issues concerning concrete and concrete repairs, part 2. *Concrete International*, 14(10), 48–52.

Schrader, E.K. (1992b) Mistakes, misconceptions, and controversial issues concerning concrete and concrete repairs, part 3. *Concrete International*, 14(11), 54–59.

Silfwerbrand, J. (1990) Improving concrete bond in repaired bridge decks. *Concrete International*, 9, 61–66.

Silfwerbrand, J. (1992) The influence of traffic-induced vibrations on the bond between old and new concrete. Bulletin No. 158, Department of Structural Mechanics and Engineering, Royal Institute of Technology, Stockholm, Sweden, 78pp.

Silfwerbrand, J. and Petersson, Ö. (1993) Thin concrete inlays on old concrete roads. *Proceedings of the Fifth International Conference on Concrete Pavement Design & Rehabilitation*, Vol. 2, April 1993, Purdue University, West Lafayette, IN, pp. 255–260.

Tabor, D. (1981) Principles of adhesion—Bonding in cement and concrete. In: *Adhesion Problems in the Recycling of Concrete* (Ed. P. Kreijger), Nato Scientific Affairs Division, Plenum, New York, NY, pp. 63–90.

Takuwa, I., Shitou, K., Kamihigashi, Y., Nakashima, H., and Yoshida, A. (2000) The application of water-jet technology to surface preparation of concrete structures. *Journal of Jet Flow Engineering*, 17(1), 29–40.

Talbot, C., Pigeon, M., Beaupré, and Morgan, D.R. (1994) Influence of surface preparation on long-term bonding of shotcrete. *ACI Materials Journal*, 91(6), 560–566.

Teodoru, G.V.M. and Herf, J. (1996) Engineering Society Cologne presents itself (NDT methods). *Proceedings of the 14th World Conference on NDT,* New Delhi, India, pp. 939–943.

Tschegg, E.K., Ingruber, M., Surberg, C.H., and Münger, F. (2000) Factors influencing fracture behavior of old-new concrete bonds. *ACI Materials Journal,* 97(4), 447–453.

Van der Wielen, A., Courard, L., and Nguyen, F. (2010) Nondestructive detection of delaminations in concrete bridge decks: A first experimental study. *Proceedings of XIII International Conference on Ground Penetrating Radar,* June 21–25, 2010, Lecce, Italy, 5p.

Zhu, Y. (1992) Effect of surface moisture condition on bond strength between new and old concrete. Bulletin No. 159, Department of Structural Mechanics and Engineering, Royal Institute of Technology, Stockholm, Sweden, 27 pp.

Chapter 8

Conclusions and perspectives

A handful of technical documents around the world address concrete surface assessment. Organizations such as the American Concrete Institute (ACI 364.1-R07, 2007), the American Standard Association, RILEM, and the European Committee for Normalization (EN 1504-10) have produced guidelines and standards intended to provide guidance to the practitioners (engineers, specifiers, QA/QC specialists, applicators, etc.) in the assessment and preparation of a concrete surface prior to the installation of repairs and surface treatments in general.

Technology is based on fundamental scientific knowledge and research developments: the theory of adhesion is the basis of all concrete surface engineering concepts. All the techniques developed for preparing concrete before application of surface treatments are directly inspired from the general principles of adhesion: creation of an interface, development of interaction, and preparation for increased contact surface.

It has been emphasized in this book that the characteristics of an existing concrete surface deeply influence the quality of surface treatments. When concrete surface treatments are not successful, the treatment material itself is often blamed for "not sticking," although the real source of trouble commonly lies in the surface preparation. Proper attention to surface preparation is essential for ensuring a durable surface treatment, notably in making sure that the contaminated and/or deteriorated concrete is completely removed and that the exposed concrete and reinforcement are sound and clean.

The concept of compatibility that has been proposed allows us to define more clearly the conditions in which adherence between a new material and existing concrete may develop and last. Deformability, chemical and electrochemical properties, permeability, and aesthetic characteristics have been shown as fundamental parameters influencing the success of the repair or surface treatment. Yet, the development of bond (chemical and physical) in the first place is a prerequisite for the concept of compatibility to materialize. This is, indeed, the ultimate objective of concrete surface engineering: to characterize and prepare the surface in such a way that an optimal bond can first develop and that compatibility of materials can be achieved.

Surface preparation prior to repair or to other concrete surface treatments is the process by which a sound, clean, and suitably roughened surface is produced, such as to provide optimal bonding conditions. This process involves the removal of all unsound and, inevitably, some sound concrete, as well as the elimination of any bond-inhibiting foreign materials from the concrete and reinforcement surfaces, the opening of the concrete pore structure, as well as reinforcement condition assessment and repair, if necessary.

Concrete removal operations have to be performed in a manner ensuring that the structural integrity of the residual concrete and reinforcing steel is maintained. Surface preparation should yield a sound, uniform, and clean surface, free of dust, dirt, and oil. After surface preparation, the concrete surface should have the following characteristics:

- General properties close to those of the nondeteriorated bulk concrete, notably for tensile strength
- Moisture content of the substrate adequate for the intended treatment
- Level of roughness adequate for the intended treatment
- Uniformity
- Absence of large voids, honeycombs, cracks, and any foreign material or object

Implementation of adequate quality control concrete surface preparation is an important consideration for the success of most surface treatments. European Standards EN 1504 series give some indication on the techniques to be used and how the quality control must be implemented (EN 1504-10). In the absence of specific instructions, surface characterization should include the following measurements:

- Superficial tensile strength (pull-off testing)
- Compressive strength assessed (*Schmidt* hammer, cores)
- Roughness
- Moisture content (carbide bomb, electrical resistance measurement)
- Water absorption (Karsten's tube)

While adequate concrete surface preparation and reliable characterization are critical to the success of concrete surface treatments, it remains difficult to simply and quickly evaluate the quality of surface. In particular, cleanliness, roughness, and saturation level are offering practical challenges. The difficulties are either related to the availability of suitable field test methods or to test result interpretation. In that perspective, simple developments such as the extension of the CSP roughness index toward the coarser range would be beneficial. Efforts must be devoted to a quantitative assessment of microcracking. New and more accurate NDT methods devised for the near-to-surface concrete could certainly be helpful for that purpose. Water content determination remains another characterization area in need for much improvement, as it can considerably influence the adherence of a given treatment. Reliable test methods that are adapted to field conditions must be developed or evaluated in order to allow adequate characterization and QC testing.

The ever-increasing diversity of advanced materials being used in new concrete infrastructure is also among the various challenges faced by the engineers. The use of specialty concretes such as high-performance concrete, ultrahigh-performance concrete, and self-compacting concrete is progressively becoming common trade. The protection and repair of structures built with these materials will certainly raise questions with regards to surface preparation and requirements for compatibility. For instance, in the case of some very high strength materials, will the current surface preparation techniques be adapted? Considering a potentially more brittle behavior of the concrete, could there be some risk of inducing more severe cracking during the surface preparation operations? Besides, many of the newest materials have significantly lower porosity than ordinary concrete. Does it mean that special surface preparation means will be necessary to yield sufficient mechanical interlocking between the existing concrete and the surface treatment? To ensure mechanical compatibility with materials exhibiting peculiar volume change characteristics, will this necessitate a more systematic use of specially adapted bonding materials?

Questions have been addressed in this book and a number of them remain partially answered or unanswered. While concrete surface engineering remains an open field for research and development, its technological advances will definitely have some economical incidence in the forthcoming decades. The needs for repair and rehabilitation of concrete infrastructure will keep expanding as an astounding number of constructions around the world are reaching more than half a century. Researchers and engineers must keep devoting efforts to solve the technological problems associated with concrete deterioration and develop optimized strategies for concrete structure maintenance.

"Who does not doubt acquires little."

Leonardo da Vinci

Index